NATIONAL GEOGRAPHIC

ANGRY BIRDS SPACE

Previous page, Earthrise over the moon, 1968

Hubble Space Telescope view of the star V838 Monocerotis

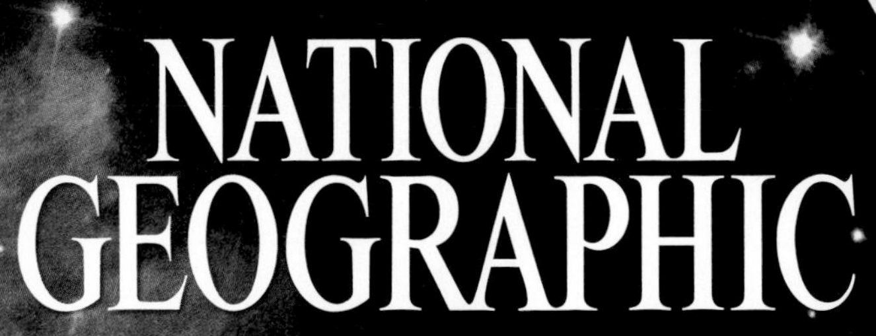

A FURIOUS FLIGHT INTO THE FINAL FRONTIER

AMY BRIGGS ★ FOREWORD BY PETER VESTERBACKA

Washington, D.C.

Published by the National Geographic Society, 1145 17th Street N.W., Washington, D.C. 20036

ISBN: 978-1-4262-0992-5

The National Geographic Society is one of the world's largest nonprofit scientific and educational organizations. Founded in 1888 to "increase and diffuse geographic knowledge," the Society's mission is to inspire people to care about the planet. It reaches more than 400 million people worldwide each month through its official journal, *National Geographic,* and other magazines; National Geographic Channel; television documentaries; music; radio; films; books; DVDs; maps; exhibitions; live events; school publishing programs; interactive media; and merchandise. National Geographic has funded more than 9,600 scientific research, conservation and exploration projects and supports an education program promoting geographic literacy.

For more information, visit **www.nationalgeographic.com.**
National Geographic Society
1145 17th Street N.W.
Washington, D.C. 20036-4688 U.S.A.

For information about special discounts for bulk purchases, please contact
National Geographic Books Special Sales: **ngspecsales@ngs.org**

For rights or permissions inquiries, please contact National Geographic Books
Subsidiary Rights: **ngbookrights@ngs.org**

Cover image: NASA/ESA and Adolf Schaller
Book design by Jonathan Halling
Printed in the United States of America
12/WOR/2

Contents

THE SKY IS NOT THE LIMIT

The stars have always inspired children and adults, scientists and explorers, as well as science-fiction writers and moviemakers. Space is an enormous place, but we have seen only a small fraction of it, firing up our imaginations with new ways to explore our solar system and search for strange new worlds. Space programs around the world have aimed to discover even more information of our surrounding universe. Our generation owes a great debt to all those brave people who had the courage to dream and take on quests that seemed impossible at the time. (Thanks to them we now know that the moon is not made of cheese!)

This outstanding book is another example of a leap into the great unknown, and the result of a successful partnership between two entities that are the leaders in their fields: Rovio, the leading entertainment company famous for the Angry Birds, and National Geographic, one of the largest nonprofit scientific and educational institutions in the world. Take on a quest of your own and explore space alongside the Angry Birds in this extraordinary book. From the first page to the last it's filled with incredible illustrations and awesome facts to fuel your imagination! This amazing book proves that one should never stop dreaming and reaching for the stars—or beyond!

Peter Vesterbacka
Mighty Eagle & CMO
Rovio Entertainment Ltd.

IN SPACE, NO ONE CAN HEAR YOU SQUEAL!

On a remote island lives a group of Angry Birds who share a common purpose: to stop green pigs from stealing and devouring their eggs. One fateful afternoon after the pigs absconded to their fortress with stolen eggs, the birds attacked, saving the eggs and demolishing the stronghold. Then, a huge vortex opened up in the sky. Out of the hole flew a strange bird shaped like an ice cube. With him was a glowing egg. Clearly they were not from this world!

As the birds and pigs stared skyward, a huge metallic arm emerged from the vortex, snatched the glowing egg, and retreated into the darkness. Evil laughter echoed over the island. The alien bird chased them back into the wormhole, which began to close. As the vortex shrank, the Angry Birds noticed that their eggs too were gone! Without a moment's hesitation they flung themselves into the wormhole just before it shut. At the end of the swirling space tunnel, the Angry Birds found themselves dressed like superheroes on a strange new planet. They saw strange piggies escaping farther into space with their eggs!

From Mercury to the Milky Way, space holds an infinite number of hiding places for the piggies. How will the Angry Birds ever save their precious eggs? By learning all they can about space in order to track down the alien pigs and rescue their eggs. Follow the Angry Birds on their latest adventure as they make new friends and gather "Space Data" and amazing "Astrofacts" about planets, stars, space exploration, galaxies, and the worlds beyond, in a daring rescue mission that will take them across the galaxy. Only then will they be able to take on the pigs, retrieve their eggs, and head back home.

LEVEL 1 YOU ARE HERE

WHAT'S OUT THERE?

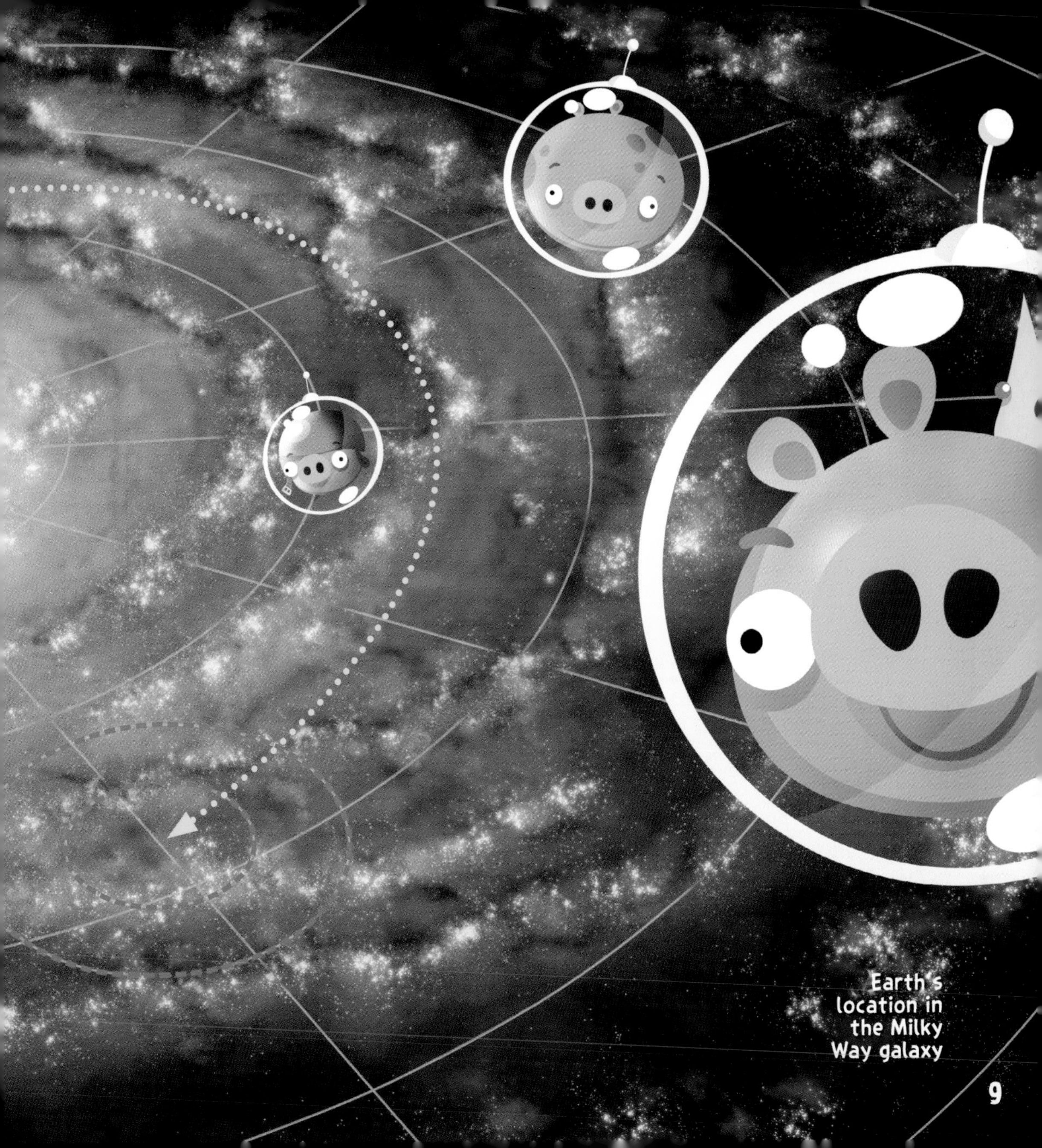

Earth's location in the Milky Way galaxy

HOME BASE

The planet Earth makes its home in a solar system located in the outskirts of the Milky Way, a spiral galaxy that contains billions of stars. The galaxy has three distinct parts. At the core is a bright, bar-shaped bulge of yellow and red stars. From the center, several spiral arms extend outward, forming a disk that contains stars of all ages and colors as well as glowing regions of star birth. Earth resides about 28,000 light-years from the galaxy's center. Surrounding the galactic disk is a great, invisible halo of dark matter that holds most of the galaxy's mass.

ASTROFACT

THE SOLAR SYSTEM ORBITS THE CENTER OF THE MILKY WAY EVERY 220 MILLION YEARS.

SPACE DATA

NAME: MILKY WAY GALAXY

SIZE: 100,000 LIGHT-YEARS ACROSS, 1,000 LIGHT-YEARS THICK

AGE: GALACTIC HALO, 13.2 BILLION YEARS

A bright Perseid meteor

THE SOLAR SYSTEM

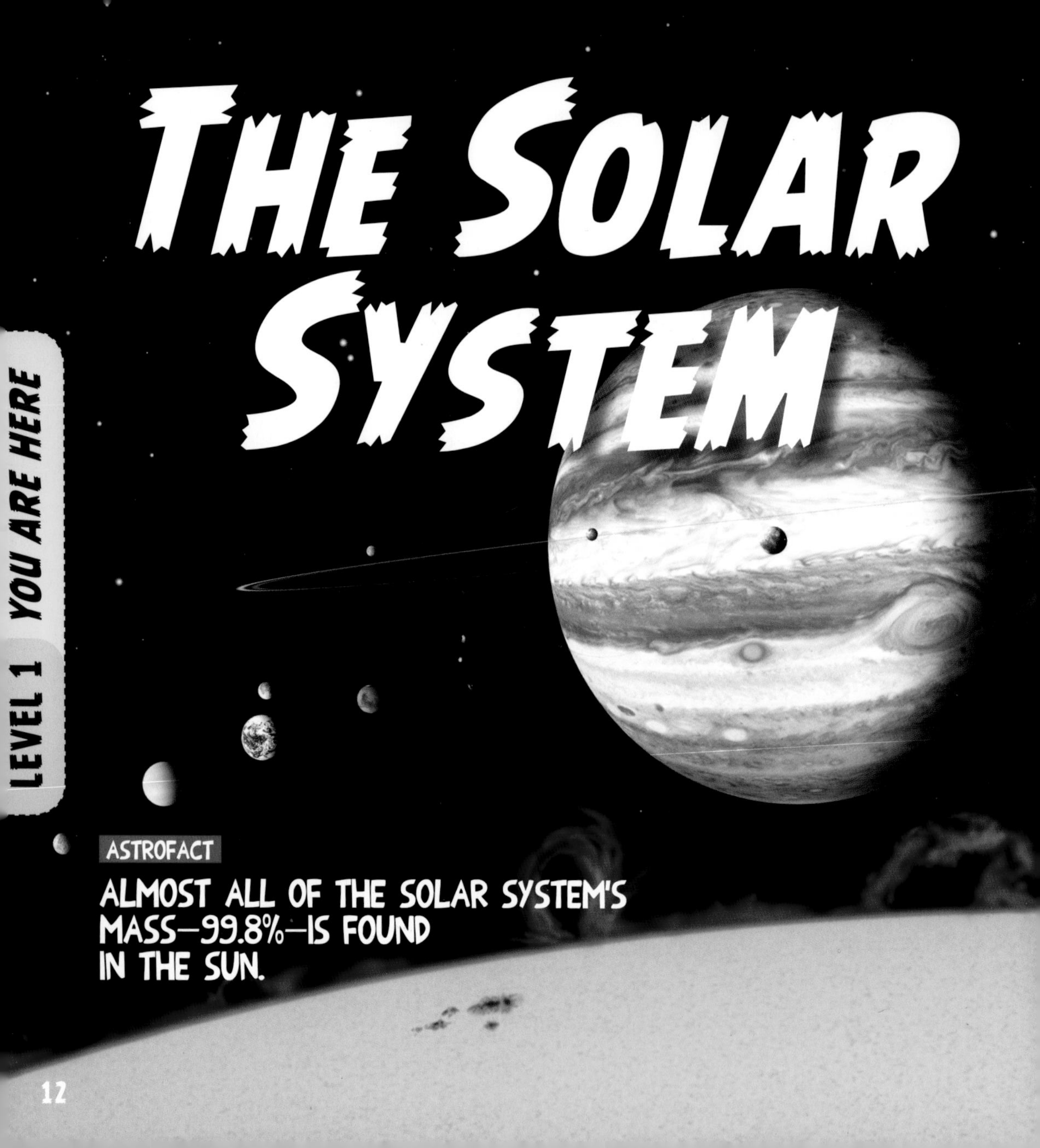

ASTROFACT

ALMOST ALL OF THE SOLAR SYSTEM'S MASS—99.8%—IS FOUND IN THE SUN.

Everything in the solar system is dominated by the sun, whose gravity holds a diverse array of celestial objects together. These range in size from planets down to asteroids and even dust, and they stretch from the sun's surface to the distant Oort cloud, home to billions of massive ice balls, which if dislodged, fall in toward the sun and turn into long-tailed comets. The planets spread across a distance ranging from Mercury, on average 36 million miles (58 million km) from the sun, to Neptune, whose orbit keeps it at an average distance of about 2.8 billion miles (4.5 billion km) from the sun.

ASTROFACT

SCIENTISTS MEASURED THE AGE OF THE SOLAR SYSTEM BY DETERMINING WHEN PRIMITIVE METEORITES FORMED: 4.57 BILLION YEARS AGO.

The solar system

WHAT'S A PLANET?

In 2006 the International Astronomical Union (IAU) announced that Pluto was being demoted. The reason? Discoveries of new objects bigger than Pluto—some with their own moons—made it necessary for astronomers to reconsider what is, or is not, a planet. So what is a planet now? The criteria: A planet must orbit a star, not be a satellite, remain spherical by the pull of its own gravity, and be big enough to clear other objects from its orbit. Pluto fails the last condition, so it was demoted to "dwarf planet." The same IAU ruling gave a name to the category of dwarf planets beyond Neptune—plutoid—which also includes Eris, Makemake, and Haumea.

ASTROFACT

CERES, ONCE JUST AN ASTEROID, WAS PROMOTED TO DWARF PLANET IN 2006.

ASTROFACT

THE BIGGEST PLANET DISCOVERED IN THE MILKY WAY IS CT CHAMAELEONTIS B, WITH MORE THAN TWICE JUPITER'S DIAMETER.

Pluto and its largest moon, Charon

ROCKET SHIPS

The idea of space travel isn't new; the concept was used in science fiction, most notably by Jules Verne, in the mid-19th century. The scientific argument for launching and using satellites began in 1903, when Russian scientist Konstantin Tsiolkovsky calculated the orbital speed required to launch an object into orbit around the Earth. In 1926 American Robert H. Goddard launched the first liquid-fueled rocket, a combination of liquid oxygen and gasoline. The rocket flew for only a few seconds, climbed 41 feet (13 m), and landed in a cabbage patch about 200 feet (60 m) away. These early efforts laid the groundwork for the technology that would take us to the moon, the planets, and beyond.

Slowpoke.

ASTROFACT

WITH TODAY'S TECHNOLOGY, IT WOULD TAKE ABOUT 18,000 YEARS FOR A SPACECRAFT TO REACH PROXIMA CENTAURI, THE NEAREST STAR.

A 1927 sketch of future rocket travel

BLAST OFF!

Humankind's first experiment with launching objects into orbit occurred more than 50 years ago when, on October 4, 1957, the Soviet Union launched Sputnik I, the first man-made satellite; the first American satellite, Explorer I, would follow on January 31, 1958. Since then, humans have launched thousands of objects into space, and many still orbit our planet. Today, satellites are used largely for global communications and for scientific research and exploration. There are communications satellites, weather satellites, navigational satellites, and astronomical satellites, like the Hubble Space Telescope, which collect data about the universe.

ASTROFACT

SPUTNIK, THE FIRST MAN-MADE SATELLITE TO ORBIT EARTH, WAS THE SIZE OF A BEACH BALL BUT WEIGHED 184 POUNDS (84 KG).

SPACE DATA

FIRST MAN-MADE SATELLITE: SPUTNIK I, OCTOBER 4, 1957

DIAMETER: 23 INCHES (58 CM)

TIME TO COMPLETE ONE ORBIT: 98 MINUTES

NUMBER OF TIMES AROUND THE EARTH: 1,440

IN 1945, SCIENCE-FICTION WRITER ARTHUR C. CLARKE PUBLISHED THE IDEA OF USING SATELLITES FOR HIGH-SPEED, GLOBAL COMMUNICATION.

Sputnik I, the first man-made satellite

ASTROFACT

THE FIRST ANIMAL TO ORBIT EARTH WAS LAIKA, A FEMALE DOG WHO TRAVELED ON SPUTNIK 2 IN 1957.

Launch of Alan Shepard's space capsule, Freedom 7, May 1961

SPACE PIONEERS

People started traveling to space shortly after Sputnik. In April 1961, Soviet cosmonaut Yuri Gagarin became the first person to orbit the Earth. Gagarin made a single orbit of the planet, exited his space capsule once it had reentered the atmosphere, parachuted back down to the ground, and, after he landed, had to find a phone to call Moscow. American Alan Shepard soon followed on May 5, 1961, with the Mercury-Redstone 3 mission. The first American to orbit the planet was John Glenn, with the Mercury-Atlas 6 mission in February 1962.

ASTROFACT

THE FIRST HUMANS TO ESCAPE EARTH'S GRAVITY WERE FRANK BORMAN, JAMES LOVELL, AND WILLIAM ANDERS, THE CREW OF APOLLO 8, IN 1968.

SPACE DATA

- FIRST PERSON IN SPACE: YURI GAGARIN, APRIL 12, 1961
- FIRST WOMAN IN SPACE: VALENTINA TERESHKOVA, JUNE 16, 1963
- YOUNGEST PERSON IN SPACE: GHERMAN TITOV, AGE 25, AUGUST 6, 1961
- OLDEST PERSON IN SPACE: JOHN GLENN, AGE 77, OCTOBER 29, 1998

TO THE MOON!

After the Soviet Union crash-landed the first space probe on the moon's surface in 1959, humans began racing for the moon. They reached it during the United States' Apollo program. Beginning on July 20, 1969, with the Apollo 11 mission, and continuing through December 19, 1972, with Apollo 17, NASA launched six missions that landed 12 people on the moon, collected a wealth of scientific data, and brought back 842 pounds (382 kg) of lunar rocks for further study. Since 1972, no person has gone to the moon, although several nations have expressed interest in returning.

ASTROFACT

ASTRONAUT NEIL ARMSTRONG LEFT HIS SPACE BOOTS ON THE MOON.

Astronaut Edwin "Buzz" Aldrin on the moon

ASTROFACT

TO BIKE TO THE MOON YOU WOULD HAVE TO PEDAL NONSTOP FOR ABOUT THREE YEARS.

Astronaut Rick Mastracchio outside the International Space Station

ASTROFACT

A SPACE SUIT WEIGHS 250 POUNDS (114 KG) WITHOUT AN ASTRONAUT IN IT.

SUITING UP

In the past space suits were custom-made for each astronaut and used just once per mission. But today, because conditions can be more easily controlled on a space station, an astronaut wears a lot of different outfits, depending on the situation. Inside the station, crew members can wear just about anything: pants, shorts, T-shirts, or sweaters. They only don a big bulky space suit, or Extravehicular Mobility Unit (EMU), when going on a space walk. Protecting them from the extreme conditions of space, the EMU also contains a life-support system, electrical power, water-cooling equipment, ventilation fan, and an in-suit drink bag.

ASTROFACT

ASTRONAUTS USE "LOWER TORSO ASSEMBLY DONNING HANDLES" TO PULL UP THEIR SPACE SUIT PANTS.

HANGING OUT

So much cushier than a slingshot.

ASTROFACT

IN NOVEMBER 2011, RED BIRD RODE A ROCKET TO THE INTERNATIONAL SPACE STATION. THE TOY BELONGED TO THE DAUGHTER OF ONE OF THE ASTRONAUTS, WHO USED HIM AS A "ZERO-G INDICATOR."

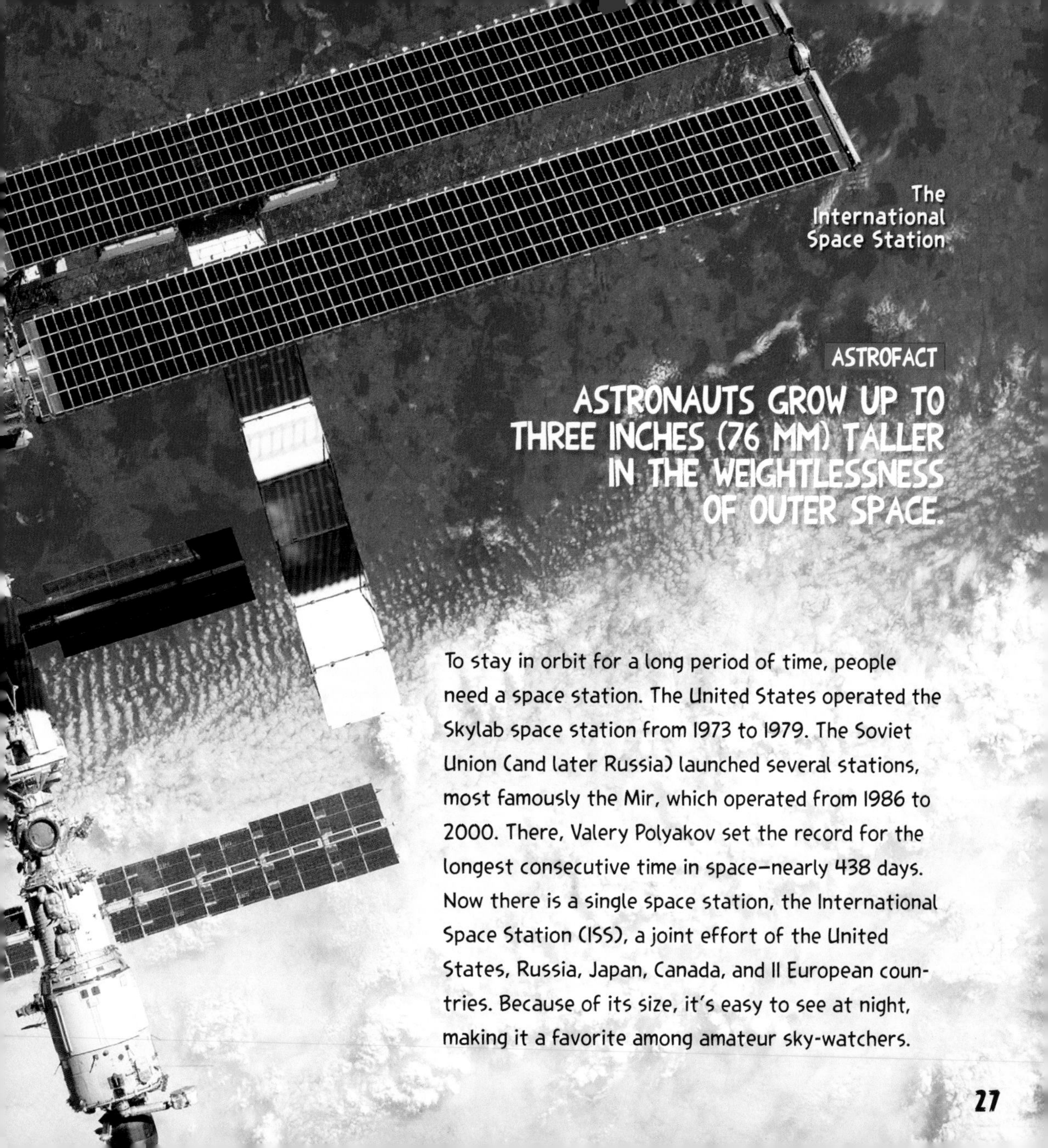

The International Space Station

ASTROFACT

ASTRONAUTS GROW UP TO THREE INCHES (76 MM) TALLER IN THE WEIGHTLESSNESS OF OUTER SPACE.

To stay in orbit for a long period of time, people need a space station. The United States operated the Skylab space station from 1973 to 1979. The Soviet Union (and later Russia) launched several stations, most famously the Mir, which operated from 1986 to 2000. There, Valery Polyakov set the record for the longest consecutive time in space—nearly 438 days. Now there is a single space station, the International Space Station (ISS), a joint effort of the United States, Russia, Japan, Canada, and 11 European countries. Because of its size, it's easy to see at night, making it a favorite among amateur sky-watchers.

ASTROFACT

A MANNED TRIP TO AND FROM MARS WOULD TAKE AS MANY AS 900 DAYS.

Mars rover Curiosity

ROBOTS ON MARS

A manned trip to Mars remains a goal of many space agencies around the world, but robotic exploration has been a big success. NASA's Mars rovers Spirit and Opportunity landed in 2004. Each the size of a small golf cart, the two discovered ancient water-altered sediments. The Phoenix lander, touching down near Mars's north pole in 2008, uncovered frozen water just inches below the surface. Next up is Curiosity, launched in 2011 and landing in 2012. Part rover–part science lab, Curiosity has a fully operational laboratory so it can perform a wide range of tests on-site.

ASTROFACT

THE MARS ROVER OPPORTUNITY HAS OUTLASTED ITS PLANNED 90-DAY MISSION BY MORE THAN 30 TIMES.

SATURN SURVEYOR

ASTROFACT

THE HUYGENS PROBE WAS THE FIRST CRAFT TO LAND ON A BODY OF THE OUTER SOLAR SYSTEM.

The Cassini orbiter

Seventeen countries collaborated in the Cassini-Huygens mission to Saturn and its moons, and they haven't been disappointed. The Cassini-Huygens mission launched in 1997 but didn't reach Saturn until 2004. On arrival, the orbiter released the Huygens probe, which parachuted to a soft landing on Saturn's moon Titan. Since then the discoveries have been many: strange weather patterns, new moons, methane lakes on Titan, and ice geysers on the moon Enceladus. Extended in 2010, Cassini's journey is now scheduled to end in 2017 when it dives into the planet for one last set of observations.

ASTROFACT

IT TOOK SEVEN YEARS FOR THE CASSINI-HUYGENS SPACECRAFT TO FLY TO SATURN.

ASTROFACT

LAUNCHED IN 1977, VOYAGER 1 IS THE FARTHEST ARTIFICIAL OBJECT IN EXISTENCE—119 TIMES FARTHER FROM THE SUN THAN EARTH.

LEVEL 1 YOU ARE HERE

ASTROFACT

AS THEIR NUCLEAR POWER SOURCES FADE, AROUND 2025, VOYAGERS 1 AND 2 WILL LOSE RADIO CONTACT WITH EARTH AND THEIR MISSIONS WILL END.

NASA's New Horizons spacecraft

Going Beyond

Bound for Pluto and beyond, NASA's New Horizons spacecraft is equipped with instruments for imaging Kuiper belt objects in visible, ultraviolet, and infrared light. The spacecraft can map their surfaces and study atmospheres. Leading the way, NASA's Voyagers 1 and 2, launched more than 30 years ago, are now headed out of the solar system on two separate trajectories. In the event they should encounter intelligent life on their long journeys, each carries a 12-inch (30 cm) gold-plated phonograph record. Drawings on the cover demonstrate how to play the record, show a map of our solar system relative to 14 pulsars (stars that emit radiation in a regular pulse-like pattern and could serve as beacons), and depict a hydrogen atom in its two lowest states, among other images.

EVERY YEAR, THE HUBBLE SPACE TELESCOPE MAKES MORE THAN 20,000 OBSERVATIONS, OR ROUGHLY 55 IMAGES EVERY DAY.

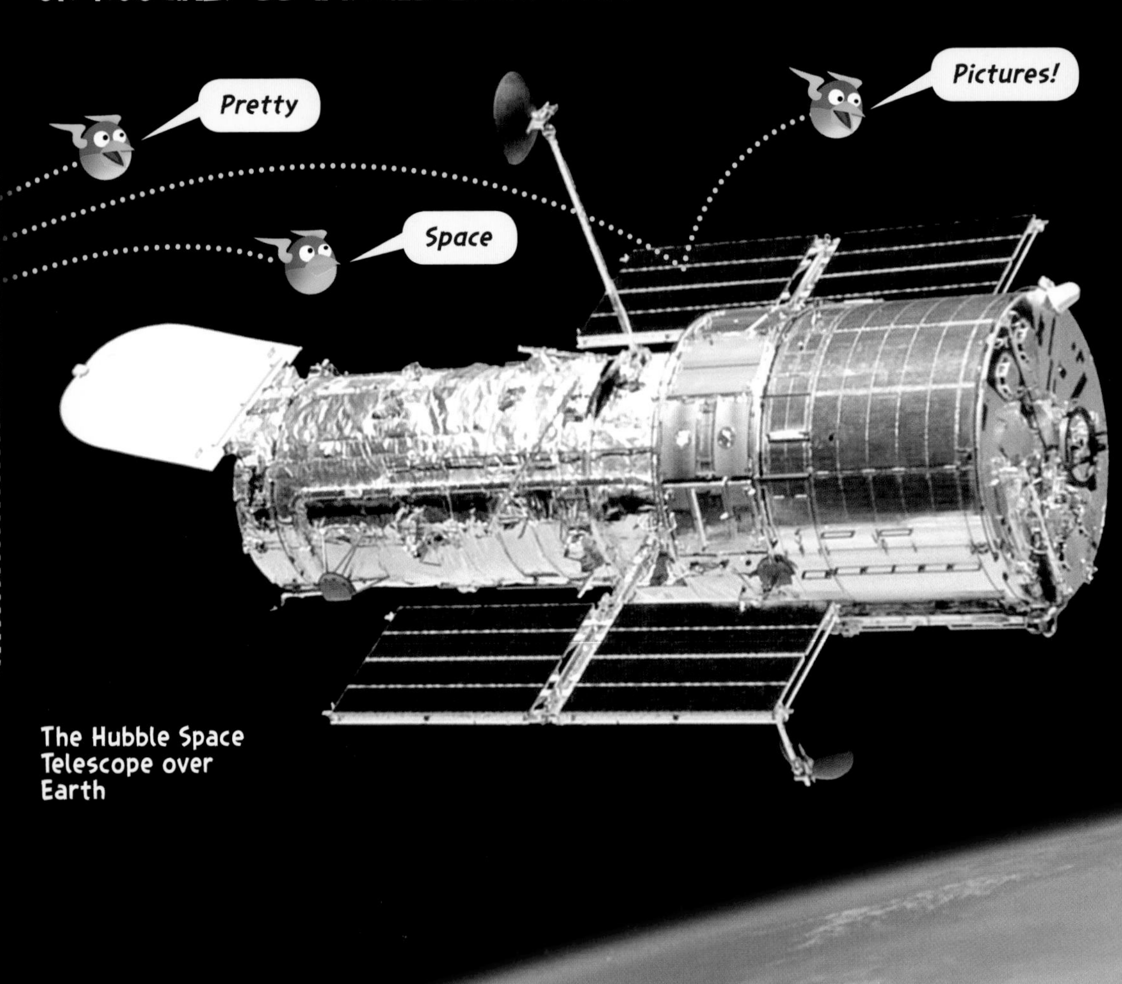

The Hubble Space Telescope over Earth

LEVEL 1 YOU ARE HERE

Looking Out

Astronomers long dreamed of a telescope that could see the cosmos unfiltered by Earth's atmosphere. That dream became reality in 1990, when the Hubble Space Telescope went into orbit. Roughly the size of a large bus, it carries a primary mirror 7.9 feet (2.4 meters) in diameter. It holds an array of instruments, including three cameras, two spectrographs, and fine-guidance sensors used to aim the telescope. Producing images ten times better than ground-based observatories, Hubble has revealed the distant realms of the universe and helped expose the existence of dark energy.

ASTROFACT

HUBBLE ORBITS SOME 380 MILES (612 KM) ABOVE THE EARTH AND TRAVELS AT 17,000 MILES AN HOUR (28,000 KM/H).

Seeing Infrared

Interstellar dust clouds and inky stretches of deep space can appear dull to ordinary telescopes, but to a telescope orbiting 26 million miles (42 million km) from Earth with the keenest infrared vision ever, they are alive with light—infrared light, or heat rays. Since its launch in 2003, the Spitzer Space Telescope has exposed formerly shrouded cosmic birthplaces of stars, planets, and galaxies. Not only is Spitzer providing information about planet formation, it is also helping astronomers understand how light and radiation from existing stars can trigger the collapse of gas clouds, forming new stars.

Nebula RCW 120, glowing in infrared colors, photographed by Spitzer

ASTROFACT

EVERY OBJECT THAT IS NOT CHILLED TO ABSOLUTE ZERO RADIATES INFRARED LIGHT.

ASTROFACT

THE SPITZER SPACE TELESCOPE IS SUPERCOOLED BY LIQUID HELIUM.

vision of
the Carina
Nebula

ASTROFACT

CHANDRA CAN SEE THE X-RAYS FROM PARTICLES UP TO THE LAST SECOND BEFORE THEY FALL INTO A BLACK HOLE.

X-RAY VISION

Named for Indian-American astrophysicist Subrahmanyan Chandrasekhar, the Chandra X-ray Observatory studies high-energy objects—such as supernovae, binary stars, and black holes—that radiate energy in x-ray wavelengths. Chandra, which was sent into orbit in July 1999, has thrilled astronomers with its discoveries about the formation of galaxies, the life cycles of stars, and the nature of black holes. The world's most powerful x-ray telescope, it can detect x-ray sources only one-twentieth as bright as earlier telescopes could. Recent observations of a vast field of intergalactic gas 400 million light-years away provide strong evidence that astronomers' long-sought "missing matter"—the huge piece of the universe that they can only theorize about and cannot even prove exists—may be hot, diffuse gas.

ASTROFACT

THE ORBIT OF THE CHANDRA X-RAY OBSERVATORY TAKES IT MORE THAN ONE-THIRD OF THE WAY TO THE MOON.

ANYBODY OUT THERE?

The Allen Telescope Array (ATA), located in Hat Creek, California, is listening for life. The ATA is a large network of 42 small dishes that hunt for radio signals bearing the mark of intelligent extraterrestrial life. Run by the SETI Institute, the ATA launched in 2007 and hopes to expand to 350 antennas. Public funding has been fitful, but private donations have helped keep the dishes online. To help keep the lights on, SETI is proposing that the ATA be used to track orbital objects that could pose threats to other satellites and prevent them from colliding.

ASTROFACT

"JINGLE BELLS" WAS THE FIRST SONG SENT BACK FROM SPACE, PLAYED ON A HARMONICA BY A CREW MEMBER OF GEMINI 6 IN 1965.

ASTROFACT

AUSTRALIA AND SOUTH AFRICA HAVE PLANS TO BUILD THEIR OWN LISTENING ARRAYS OF TELESCOPES.

The Allen Telescope Array, Hat Creek, California

LEVEL 2 SHORT FLIGHTS

The Inner Solar System

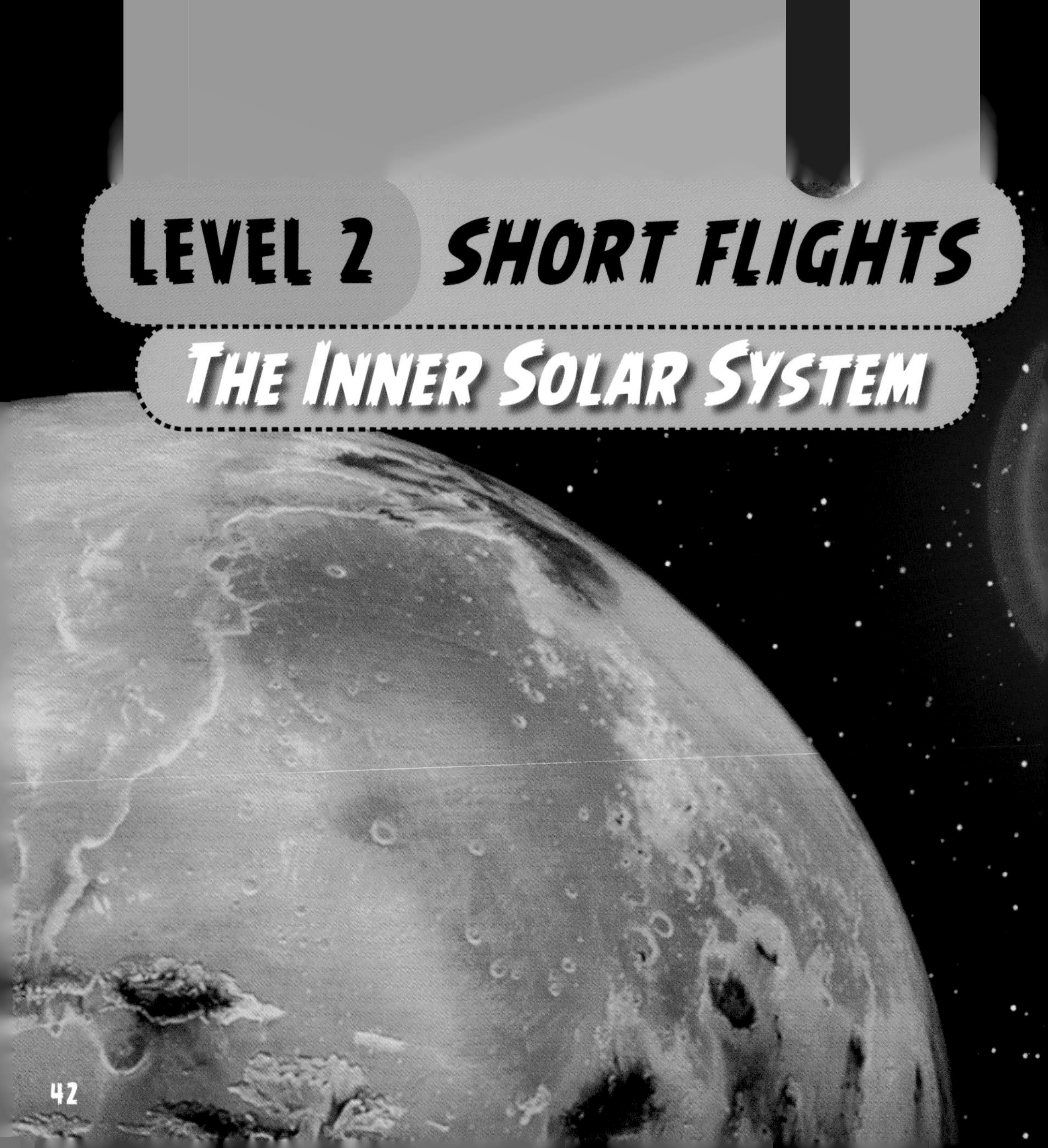

Mercury, Venus, Earth, and Mars (clockwise from upper left)

THE SUN

Born roughly 4.6 billion years ago, the sun is the closest star to Earth. A modest-size star, this hot ball of glowing ionized gases is made of mostly hydrogen and helium (plus trace amounts of such elements as oxygen, carbon, iron, and sulfur) and accounts for more than 99.8% of the entire solar system's mass. At the sun's center, the fantastic heat and pressure cause nuclear fusion to convert half a billion tons of hydrogen into helium every second. At the core, temperatures reach around 27,000,000°F (15,000,000°C), while at the sun's visible surface, it's a relatively chilly 10,000°F (5500°C).

ASTROFACT

NUCLEAR FUSION IN THE SUN'S CORE GENERATES ABOUT 400 TRILLION TRILLION WATTS OF ENERGY PER SECOND.

The sun as seen by a solar observatory

THE SUN HAS ENOUGH ENERGY TO BURN FOR 100 BILLION MORE YEARS, BUT MOST OF THAT TIME IT WOULD BE A COOLING, WHITE DWARF.

Watch out for burn-out!

SPACE DATA

- DISTANCE FROM EARTH: 93 MILLION MILES (150 MILLION KM)
- ROTATION PERIOD: 25.38 EARTH DAYS
- SURFACE TEMPERATURE: 9939°F (5504°C)
- DIAMETER: 864,000 MILES (1.39 MILLION KM)
- MAJOR PLANETS: 8

A MIGHTY WIND

The sun dominates the solar system not only through its gravitational influence, which extends to the Oort cloud as far away as 200,000 AU (astronomical units—one AU is the mean distance between the Earth and sun), but also through its solar wind of charged particles that reaches well past Pluto's orbit. The solar wind involves a fairly constant stream of charged particles from the sun—roughly two million tons of matter per second streaming at speeds as fast as two million miles an hour (three million km/h). The solar wind creates the heliosphere, a vast bubble in the interstellar medium that surrounds the solar system.

ASTROFACT

EARTH'S MAGNETIC FIELD PROTECTS IT FROM THE SOLAR WIND BY DEFLECTING MOST OF THE CHARGED PARTICLES.

THE RADIATION FROM SOLAR FLARES COULD BE DEADLY TO ASTRONAUTS CAUGHT OUTSIDE THEIR SPACECRAFT.

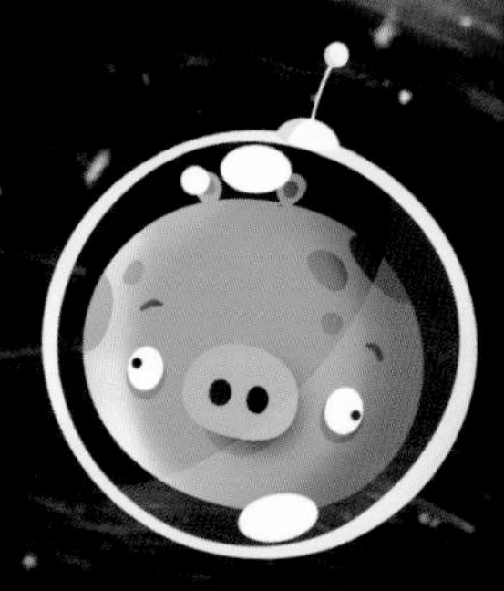

The heliosphere (blue), a bubble in which the sun and planets reside, inflated by the solar wind

Seeing Spots

Typically appearing near the sun's equator, sunspots peak and ebb in an 11-year cycle that coincides with other solar activity. The 11-year cycle is part of a longer 22-year period in which the magnetic fields of spots in the sun's upper and lower hemispheres switch polarities. Solar activity near sunspots can build up magnetic tension, which eventually gives way, and its release can eject billions of tons of atomic particles in a solar flare. These heavier than usual doses of radiation cause auroras that can be seen on Earth—and they can knock out communications equipment and satellites and disrupt the flow of electricity in terrestrial power grids. Serious stuff!

ASTROFACT

ASTRONOMERS HAVE RECORDED THE APPEARANCE OF SUNSPOTS AS FAR BACK AS 28 B.C., WHEN CHINESE ASTRONOMERS MADE NOTE OF THEM.

ASTROFACT

SUNSPOTS COME IN MAGNETICALLY LINKED PAIRS.

The dark heart of a sunspot

ASTROFACT

SCIENTISTS DON'T UNDERSTAND EXACTLY HOW THE SOLAR CORONA CAN REACH TEMPERATURES IN THE MILLIONS OF DEGREES.

Coronal loops

The Corona

The sun's outermost layer, the far-reaching corona, extends for millions of miles—a ghostly halo of gas, invisible except during total solar eclipses. Compared with the dense solar core, the corona is almost nonexistent—trillions of times less dense than the air on Earth. But temperatures in the corona inexplicably rise to perhaps 2,000,000°F (1,100,000°C), and no one quite knows why. Holes in the corona, caused by the sun's magnetic field, are where streams of particles in the solar wind break free and surge outward into the solar system. Some believe the magnetic field may be responsible for the corona's unusually high temperatures.

ASTROFACT

SOARING STRUCTURES CALLED CORONAL LOOPS ARC HIGH ABOVE THE SUN'S SURFACE—SOME REACHING AS HIGH AS TEN EARTHS.

MERCURY

The nearest planet to the sun, Mercury is just 3,031 miles (4,878 km) in diameter, making it the smallest major planet in the solar system, only slightly larger than Earth's moon. Mercury's orbit is highly eccentric, taking it as far away from the sun as 43 million miles (69 million km) and as close as 29 million miles (46 million km). With a core made mostly of iron, Mercury is the second densest planet (Earth is the first). NASA's MESSENGER mission, first launched in 2004, is studying new details about Mercury's composition and its craggy surface. In 2011, MESSENGER became the first spacecraft to orbit the planet.

ASTROFACT

MERCURY ORBITS THE SUN FASTER THAN ANY OTHER PLANET: ONE YEAR LASTS JUST 88 EARTH DAYS.

The MESSENGER spacecraft above Mercury

ASTROFACT

DAYS ARE LONGER THAN YEARS ON THE PLANET MERCURY.

SPACE DATA

- **DISTANCE FROM THE SUN:** 29–43 MILLION MILES (46–69 MILLION KM)
- **ROTATION PERIOD:** 58.6 EARTH DAYS
- **ONE YEAR:** 88 EARTH DAYS
- **DIAMETER:** 3,031 MILES (4,878 KM)
- **MOONS:** 0

THE BIGGEST IMPACT BASIN ON MERCURY IS 960 MILES (1,550 KM) WIDE AND WAS PROBABLY CREATED WHEN A HUGE ASTEROID SLAMMED INTO THE PLANET WITH A FORCE OF A TRILLION HYDROGEN BOMBS.

COVERED IN CRATERS

In 2008, NASA's probe MESSENGER reached Mercury and began visually mapping its surface, revealing details that confirm scientists' earlier beliefs. The spacecraft's first flybys revealed the prevalence of impact craters, evidence of the violent collisions that were common among all the inner planets. The detail returned by MESSENGER, however, allowed a refined view of the planet—a view so clear that the International Astronomical Union has approved titles for 291 craters. That includes the giant, 430-mile-diameter (692 km) Rembrandt Basin, thought to have formed about 3.9 billion years ago.

ASTROFACT

IN ITS EARLY HISTORY, MERCURY SHRANK AS IT COOLED, WRINKLING ITS CRUST.

A false-color image of Caloris Basin, Mercury's largest crater

VENUS

Earth's closest neighbor, Venus, comes within 24 million miles (38 million km) at its closest Earthly approach every 19 months. At first glance, Venus and Earth have a lot in common. The two planets are of similar size and structure (Earth is just a little bit bigger and heavier). Both planets have substantial atmospheres. But that's where the similarity ends. Earth's atmosphere is mostly nitrogen and oxygen, but Venus's atmosphere is made of carbon dioxide gas and sulfuric acid clouds that trap the sun's heat, causing surface temperatures to rise to inhospitable highs of 860°F (460°C).

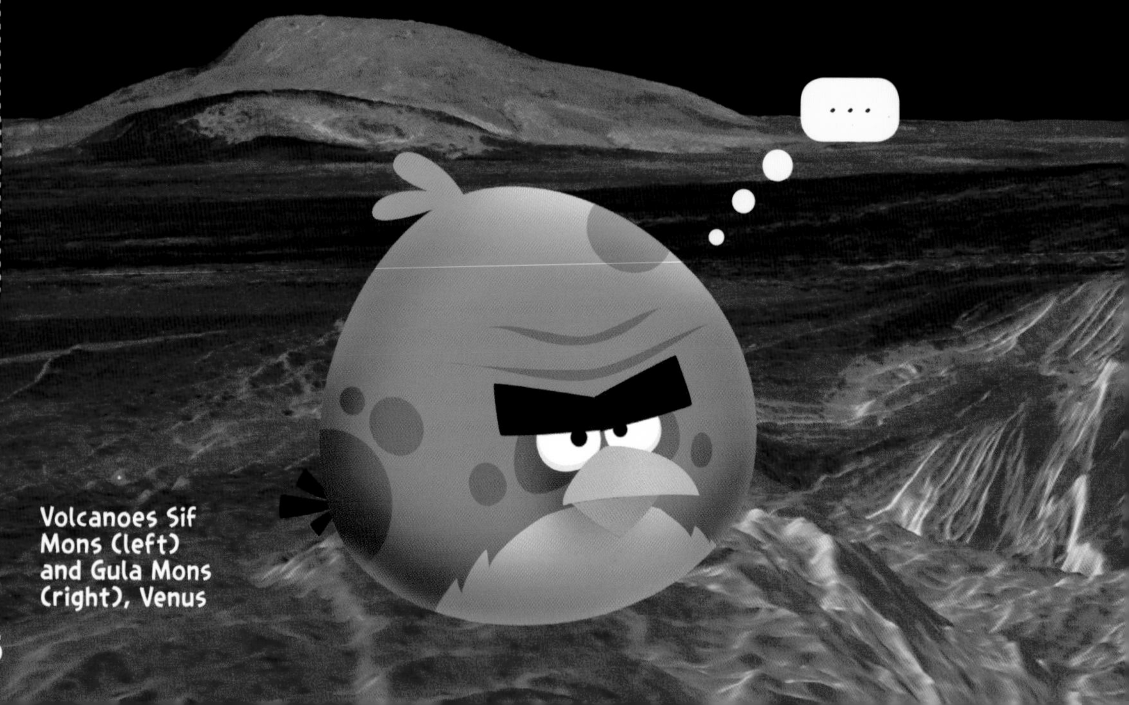

Volcanoes Sif Mons (left) and Gula Mons (right), Venus

ASTROFACT

VENUS NEVER COOLS OFF AFTER DARK—MIDNIGHT IS AS HOT AS NOON: 860°F (460°C).

ASTROFACT

VENUS HAS NO MOONS (AND NEITHER DOES MERCURY).

SPACE DATA

- DISTANCE FROM THE SUN: 67.2 MILLION MILES (108.2 MILLION KM)
- ROTATION PERIOD: 243 EARTH DAYS
- ONE YEAR: 225 EARTH DAYS
- DIAMETER: 7,521 MILES (12,103 KM)
- MOONS: 0

Venus alongside the moon

THE EVENING STAR

Beautiful, mysterious Venus is the brightest object in the night sky after the moon. It's known both as the morning star and the evening star because Venus is visible for a few months each year either after sunset near the western horizon or before sunrise in the east. The planet's nighttime brilliance is the result of two factors: its nearness to Earth and its highly reflective sulfuric dioxide clouds, which make it the brightest of all the planets but hide its surface from view. In fact 65 percent of the sunlight that reaches it is bounced back into space.

ASTROFACT

MESOPOTAMIANS RECORDED APPEARANCES OF VENUS AS EARLY AS 3000 B.C.

ASTROFACT

PEOPLE REPORT THE MOST UFO SIGHTINGS WHEN VENUS IS CLOSEST TO EARTH.

EARTH

The Earth is unique—it's the only known place to harbor life. It is the largest and densest of the four inner, or "terrestrial," planets—Mercury, Venus, Earth, and Mars. Roughly 70 percent of its surface is covered by a liquid-water ocean. Consisting mostly of nitrogen and oxygen, Earth's atmosphere protects it from harmful radiation as well as from objects, like meteors, that might otherwise crash into the planet's surface. A magnetic field, the magnetosphere, deflects particles from the solar wind. All this protection, plus the perfect distance from the sun, has allowed life to evolve and thrive on Earth.

ASTROFACT

THE EARTH IS SLIGHTLY PEAR-SHAPED.

Me too!

SPACE DATA

DISTANCE FROM THE SUN: 93 MILLION MILES (150 MILLION KM)

ROTATION PERIOD: 24 HOURS

ONE YEAR: 365 DAYS

DIAMETER: 7,926 MILES (12,756 KM)

MOONS: 1

ASTROFACT

EARTH IS THE ONLY PLANET IN THE SOLAR SYSTEM WHERE WATER CAN EXIST AS A LIQUID, SOLID, AND GAS.

The Earth, moon, and sun in alignment

ALIEN EARTH

Exploration here on planet Earth has revealed a growing number of organisms that can survive in most unexpected places. These creatures, called "extremophiles," shun the oxygen-driven or photosynthetic processes that dominate life as we know it and thrive in some of the harshest environments. For example, life has been found around hydrothermal vents, cracks in the planet's surface where geothermally heated, scalding water bubbles through. Cracks like these may exist on the floor of the subsurface ocean on Jupiter's moon Europa. If life exists around these vents on Earth, could similar life-forms exist on Europa and elsewhere?

ASTROFACT

MICROBES CAN LIVE DEEP WITHIN RADIOACTIVE MINES AND IN THE ICE OF ANTARCTICA.

ASTROFACT

IN THE BOILING HOT SPRINGS OF YELLOWSTONE NATIONAL PARK LIVE MICROORGANISMS THAT THRIVE IN THE SCALDING WATER, STRONGER THAN BATTERY ACID.

Deep-sea hydrothermal vents, Pacific Ocean

EARTHLY VISITORS

ASTROFACT

AT LEAST 34 ROCKS FROM MARS HAVE LANDED ON EARTH AS METEORITES.

Uncountable bits of space dust, debris, and small pieces of rock orbit the sun. If one of these collides with Earth, it becomes a meteor when it enters Earth's atmosphere. Meteors reach speeds of anywhere from 20,000 to more than 160,000 miles an hour (32,000 to more than 257,000 km/h). Friction from the atmosphere heats them to perhaps 2000°F (1100°C), which causes them to vaporize in the air, leaving a bright streak, especially visible in the night sky. If the space rock reaches Earth's surface, it becomes a meteorite. The biggest and rarest ones have caused mass extinctions of life. The most prevalent meteorites are mostly silicates and other rocky material.

ASTROFACT

METEORITES THE SIZE OF BASKETBALLS LAND ON EARTH ABOUT ONCE A MONTH.

Meteor Crater, Arizona

THE MOON

Although it may look simple, the partnership between the moon and the Earth is complex. The moon spins on its axis, just as Earth does, but the moon's rate of rotation matches the rate of its progress around Earth—it takes about 27.3 Earth days to complete both. That means only one side of the moon is visible, while the other side faces away. It takes a bit longer, about 29.5 days, to complete what is known as the moon's synodic month. This is the time it takes to orbit Earth once and return to the same position relative to the sun—in other words, to go from one full moon to the next.

Maybe Earth is angry. My temperature rises when I get angry.

ASTROFACT

THE EARTH'S TEMPERATURE RISES SLIGHTLY DURING A FULL MOON.

SPACE DATA

DISTANCE FROM EARTH: 238,855 MILES (384,400 KM)

DIAMETER: 2,159 MILES (3,475 KM)

ORBITAL PERIOD: 27.3 DAYS

CYCLE OF PHASES: 29.5 DAYS

ORBITAL CIRCUMFERENCE: 1,423,000 MILES (2,290,000 KM)

The battle-scarred moon

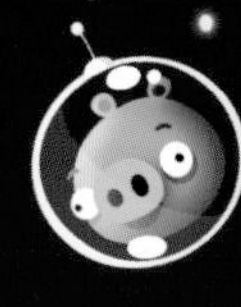

ASTROFACT

THE MOON HAS NO SEASONS. YEAR-ROUND, SUNLIGHT FALLS ALMOST DIRECTLY ONTO THE MOON'S EQUATOR, HALFWAY BETWEEN ITS TWO POLES.

ASTROFACT

DURING A TOTAL LUNAR ECLIPSE, THE DARKENED MOON TURNS A RUSTY OR BLOOD-RED HUE.

Red looks good on you!

ASTROFACT

LUNAR ECLIPSES CAN LAST AS LONG AS AN HOUR OR PASS IN A FEW MINUTES.

A ruddy lunar eclipse

Hide and Seek

An eclipse is a breathtaking, spooky sight. An eclipse happens when either the moon or the Earth blocks the light from the sun. In a lunar eclipse, the Earth lies directly between the sun and the moon, so that the Earth's shadow falls on the moon and turns the moon a ruddy color. This, of course, can happen only during a full moon. In a solar eclipse, the roles are reversed: The moon comes between the sun and the Earth, and the shadow of the moon falls on the Earth. This can happen only during a new moon.

MOON ORIGINS

Where did the moon come from? One theory suggests that the moon formed as a separate object and became trapped by Earth's gravitational pull. Evidence from lunar missions, however, supports another idea: The moon formed from the debris ejected when a massive object about the size of Mars struck the young Earth. That would account for the moon having rocks similar to those found near the surface of the Earth, the lack of water (vaporized during the explosion, though trace amounts have been found in some lunar rocks), and aspects of its orbit (the spin of the two bodies is hard to square with other lunar-birth theories).

ASTROFACT

AN EARLY MOON ORIGIN THEORY ARGUED THAT AFTER THE YOUNG EARTH FORMED, THE MOON COALESCED FROM THE LEFTOVERS.

ASTROFACT

SOME ASTRONOMERS BELIEVE THAT EARTH ONCE HAD TWO MOONS.

Collision between Earth and a Mars-size object

ASTROFACT

SHEETS OF ICE DETECTED ON THE MOON'S NORTH POLE WERE A COUPLE YARDS (METERS) THICK.

Artist's rendering of the LCROSS probe and Centaur booster en route to the moon

MOON MOISTURE

In October 2009, NASA's Lunar Crater Observation and Sensing Satellite (LCROSS) sped toward the moon's shadowed Cabeus crater, located close to the moon's south pole. Preceding the satellite by about four minutes, the probe's spent Centaur rocket casing crashed into the lunar surface, sending up clouds of debris for the satellite to analyze as it flew through. Scientists examining the data detected water. Radar analysis of the moon's northern craters conducted by India's Chandrayaan-1 lunar orbiter also turned up telltale signs of ice. The moon might have appeared arid, but all signs point to an icier existence than had been imagined previously.

ASTROFACT

THE NEWLY DISCOVERED PRESENCE OF WATER HAS INSPIRED DISTANT PLANS TO ESTABLISH A PERMANENT BASE ON THE MOON.

MARS

Three times a century, Earth and Mars share an extra-close encounter as they come within 35 million miles (56 million km) of each other. In some ways Mars is very similar to Earth. It rotates on its axis every 24.6 hours, resulting in a day much like ours. It has similar seasons. Mars has its own atmosphere, clouds, and polar caps. Mars is much smaller than Earth, however. More than a dozen flybys, orbiters, landers, and rovers, and many Earth-bound instruments, have scanned the Martian surface. Thanks to them, we know the surface of Mars better than that of any other planet.

ASTROFACT

MARS HAS TWO SMALL MOONS—PHOBOS AND DEIMOS—WHOSE NAMES MEAN "FEAR" AND "PANIC."

SPACE DATA

DISTANCE FROM THE SUN: 141.6 MILLION MILES (227.9 MILLION KM)

ROTATION PERIOD: 24.6 HOURS

ONE YEAR: 686.9 EARTH DAYS

DIAMETER: 4,222 MILES (6,794 KM)

MOONS: 2

ASTROFACT

ANCIENT BABYLONIANS, GREEKS, AND ROMANS ALL ASSOCIATED MARS WITH WAR BECAUSE OF ITS BLOOD-RED COLOR.

DISTANT PAST

ASTROFACT

THE TALLEST VOLCANO ON MARS IS ABOUT 14 MILES (22 KM) HIGH. THAT'S 68 TIMES TALLER THAN THE EIFFEL TOWER.

Scientists have confirmed the presence of frozen water on Mars, but they still want to know if liquid water once flowed there. To find answers, they are turning to Earth's own geology as well as the clues hidden on the Martian surface. Mars rover Opportunity studied the built-up layers of rock in the Victoria Crater and found evidence of water's activity. Detailed images taken from orbiting spacecraft have revealed patterns resembling the dry riverbeds and ancient floodplains on Earth. This geologic evidence supports the theory that Mars might have been a warmer, wetter place than it is today.

ASTROFACT

AN ANCIENT OUTCROP OF CARBONATE ROCK DISCOVERED BY MARS ROVER SPIRIT INDICATES A LONG-AGO PRESENCE OF WATER.

The rim of Victoria Crater on Mars

WHERE'S THE WATER?

Mars does have water, lots of it—in its two ice caps, frozen into the soil at latitudes higher than about 50° in both hemispheres, and likely abundant elsewhere beneath the surface as permafrost. In summer 2008, NASA's Mars Phoenix lander dug a shallow trench at its landing site in the high northern latitudes, revealing a shiny patch of ice that quickly evaporated. Samples scooped up and analyzed by the lander's instruments confirmed that the ice was in fact H_2O. Some scientists believe a vast ancient ocean or several gigantic lakes once existed on Mars.

ASTROFACT

IT SNOWS ON MARS, BUT THE FLAKES EVAPORATE BEFORE THEY REACH THE GROUND.

ASTROFACT

ANCIENT RIPPLES ON MARTIAN ROCKS INDICATE THAT THEY FORMED IN WATER ABOUT A FOOT DEEP.

Sharp-edged gullies suggest that water once flowed on Mars.

IN 1906, AN AMERICAN ASTRONOMER THEORIZED THAT AN ANCIENT CIVILIZATION BUILT CANALS ON MARS BEFORE DYING OUT.

Imagined Martian spaceships (left) and a Martian (right)

LEVEL 2 SHORT FLIGHTS

LIFE ON MARS?

Mars holds a special place in the human imagination. We know now that there aren't any "little green men" on the red planet, but recent discoveries have led to more questions about life on Mars. Ground-based observers detected methane on Mars, which is significant because on its own, the gas should decay within a couple centuries. Scientists reason that there must be an active source on the planet to replenish the gas, which could point in one of two directions: Either an active geothermal process is generating methane and venting it at the surface, or some sort of biologic activity is occurring beneath the surface.

ASTROFACT

COSMIC RADIATION FROM THE SUN, STARS, AND THE GALAXY WOULD KILL ANY LIFE IN THE TOP FEW YARDS (METERS) OF SOIL ON MARS.

SPACE ROCKS

In the asteroid belt, millions of rocky objects orbit between Mars and Jupiter. During the formation of the solar system, these bodies fell under the sway of Jupiter's massive gravity and never coalesced into planets themselves. Studying these asteroids can reveal much about how the planets were born and how planets and smaller bodies may have migrated as the solar system evolved. Objects in the asteroid belt frequently collide, and debris flies toward Earth. Most objects burn up in Earth's atmosphere as meteors, but others reach the planet's surface as meteorites.

ASTROFACT

SINCE 1801, MORE THAN 300,000 ASTEROIDS HAVE BEEN DISCOVERED AND OFFICIALLY NUMBERED.

ASTROFACT

ASTEROID NAMES INCLUDE ZAPPAFRANK, JABBERWOCK, AND DIORETSA (A BACKWARD-ORBITING ASTEROID).

Imaginary view from the asteroid belt

CERES

In 2006, when astronomers were forced to say just what defined a planet, Ceres found itself boosted into the new dwarf planet category. Massive enough to maintain a spherical shape, Ceres is by far the largest body in the asteroid belt, with a diameter of 580 miles (933 km). But Ceres is in a class by itself; none of the asteroids are even close to its size. The next biggest objects, Pallas and Vesta, are less than half as big. Like Earth, Ceres's interior probably has several layers, with a dense rocky core surrounded by an icy mantle. Evidence suggests that Ceres may even have a weak atmosphere.

SPACE DATA

DISTANCE FROM THE SUN: 257 MILLION MILES (413.6 MILLION KM)

ROTATION PERIOD: 9 HOURS

ONE YEAR: 4.6 EARTH YEARS

DIAMETER: 591.8 MILES (952.4 KM)

MOONS: 0

SPOTTED IN 1801, CERES WAS THE FIRST ASTEROID TO BE DISCOVERED.

ASTROFACT

IN 2007, NASA LAUNCHED THE DAWN MISSION, ON SCHEDULE TO ORBIT CERES IN EARLY 2015.

Artist's depiction of a visit to the dwarf planet Ceres

Doomsday Rock

A look at a potential asteroid impact

Asteroids and comets that pass near Earth are called near-Earth objects (NEOs). Many miss our planet, but some don't. Earth may not be as pock-marked as Mercury and the moon, but it has hidden scars, covered up by Earth's ever changing surface. Researchers have identified 175 craters and larger impact basins associated with comets, asteroids, and meteoroids. The 110-mile (177 km) Chicxulub crater, found on Mexico's Yucatán Peninsula in the 1970s, shows the geologic markings of a massive asteroid strike. The object is estimated to have been 6 to 12 miles (10 to 19 km) wide and has been dated to the end of the Cretaceous period, consistent with the disappearance of many Earth species 65.5 million years ago.

IN 1908, AN ASTEROID BLEW UP IN THE SKY OVER TUNGUSKA, SIBERIA, WITH THE FORCE OF 185 ATOMIC BOMBS.

Bet that woulda dented a few pig lairs!

ASTROFACT

NASA'S NEAR-EARTH OBJECT PROGRAM IDENTIFIES AND TRACKS OBJECTS THAT MIGHT BE ON A COLLISION COURSE WITH EARTH.

LEVEL 3 LONGER JOURNEY

THE OUTER SOLAR SYSTEM

Saturn eclipsing the sun

Jupiter

...

ASTROFACT

IF JUPITER HAD BEEN ABOUT 80 TIMES MORE MASSIVE, IT WOULD HAVE BECOME A STAR.

JUPITER

Able to fit 1,400 Earths inside it, Jupiter is the largest planet in the solar system. It is the closest gas giant to the sun, and its exceptionally wide orbit—nearly 500 million miles (800 million km)—takes almost 12 Earth years to complete. The planet is made mostly of hydrogen and helium but has a rocky core consisting of heavier elements. Jupiter contains more than twice the mass of all the other planets combined; its great gravitational pull is second only to that of the sun. But despite this massive bulk, Jupiter spins so quickly on its axis that one day lasts only ten Earth hours.

ASTROFACT

JUPITER'S GRAVITY IS SO STRONG THAT IT CAN KIDNAP COMETS FROM THEIR ORBITS.

SPACE DATA

DISTANCE FROM THE SUN: 483.7 MILLION MILES (778.4 MILLION KM)

ROTATION PERIOD: 10 HOURS

ONE YEAR: 11.9 EARTH YEARS

DIAMETER: 88,846 MILES (142,983 KM)

MOONS: 65 (51 NAMED)

Swirls & Storms

Jupiter's atmosphere puts on a real show. Its quick ten-hour rotation gives it a lava lamp appearance, as swirling wind creates bands of color in the planet's thick mix of hydrogen, helium, methane, and ammonia. Persistent storm systems add to the spectacle: The most famous is the Great Red Spot, a high-pressure zone about twice as big as Earth. Red Spot Jr., another, smaller storm system, joined the party in 2000 as a white storm and developed a red hue in 2005. Then in May 2008, Baby Red Spot appeared, but the Great Red Spot devoured it just a few months later.

ASTROFACT

JUPITER'S COLORFUL RUST-COLORED STRIPES ARE MADE OF AMMONIA HYDROSULFIDE—WHICH SMELLS LIKE ROTTEN EGGS.

ASTROFACT

JUPITER'S GREAT RED SPOT IS A STORM THAT HAS LASTED FOR HUNDREDS OF YEARS.

Jupiter's three red spots, May 2008

BABY RED SPOT

GREAT RED SPOT

RED SPOT JR.

Explosive!

ASTROFACT

MANY OF THE 150 VOLCANOES ON IO ERUPT ALMOST CONTINUOUSLY.

GIANT MOONS

Orbiting Jupiter are at least 65 moons, which interest scientists almost as much as their parent planet. The four largest—discovered by Galileo Galilei in 1610 and therefore known as the Galilean moons—might qualify as planets themselves if they orbited the sun. Enormous Ganymede, cratered Callisto, icy Europa, and volcanic Io all fascinate. Not only the largest moon in the solar system, Ganymede is also bigger than the planet Mercury and little Pluto. Io is the most geologically active body in the solar system, with volcanoes spewing plumes that reach 190 miles (300 km) above the surface.

ASTROFACT

JUPITER'S FOUR LARGEST MOONS—IO, EUROPA, GANYMEDE, AND CALLISTO—CAN BE VIEWED FROM EARTH WITH JUST A PAIR OF BINOCULARS.

A composite of Jupiter and its four largest moons

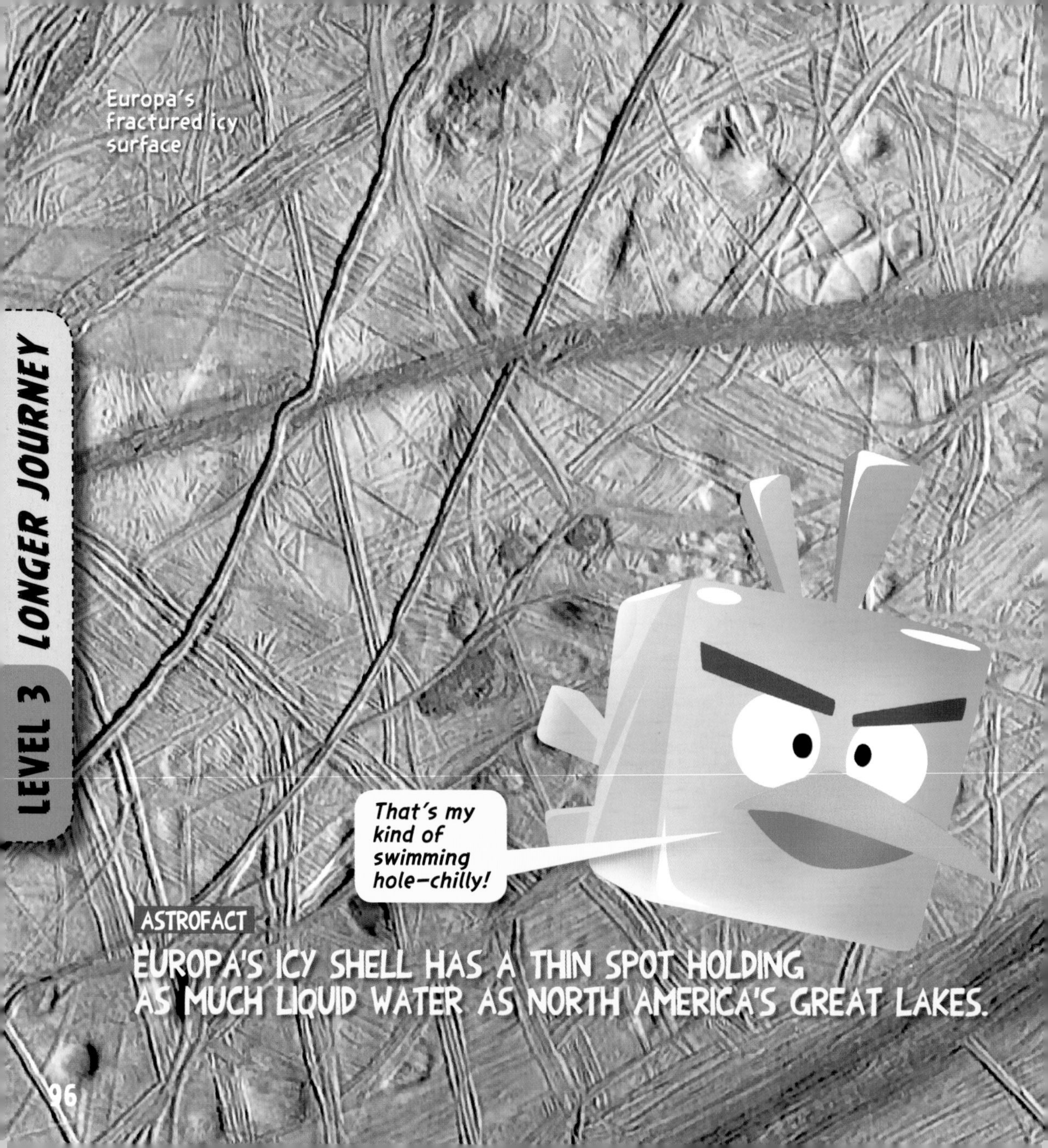
Europa's fractured icy surface
That's my kind of swimming hole—chilly!
ASTROFACT
EUROPA'S ICY SHELL HAS A THIN SPOT HOLDING AS MUCH LIQUID WATER AS NORTH AMERICA'S GREAT LAKES.

Lakes on Europa

Some scientists suspect that life could exist in an ocean on Jupiter's moon Europa. Europa has an iron core, a rocky mantle, and a fractured, thin surface layer of water ice. Under the ice lies an ocean of salty liquid water about 62 miles (100 km) deep. Images taken by NASA's Voyager and Galileo missions show the icy surface crisscrossed with fractures and chunks that resemble icebergs. Similar to formations seen on Earth, these scars on Europa may have formed when liquid below the surface welled up to fill in fissures and cracks in the ice. Recent computer models suggest that the Europan ocean could be infused with oxygen, making it a prime target in the search for life.

ASTROFACT

ASTROBIOLOGISTS WANT TO SEND A SUBMERSIBLE CRAFT TO EUROPA TO DRILL—OR MELT—THROUGH THE ICE SHELL AND PROWL EUROPA'S OCEAN.

SATURN

The most distant planet you can see with the naked eye, ringed Saturn is probably the most recognizable planet of all. This gas giant, nearly 885.9 million miles (1.43 billion km) from the sun, is far more distant than Jupiter. Its iconic rings, made of rubble, dust, and ice, stretch 170,000 miles (274,000 km) from side to side. Ammonia in its windy atmosphere gives Saturn its golden hue. The second largest planet, Saturn could contain 763 Earths, but its density is less than that of water. If you could place Saturn in a body of water, it would float.

ASTROFACT

ANCIENT MESOPOTAMIAN ASTRONOMERS CALLED SLOW-MOVING SATURN "THE OLD SHEEP."

I prefer "The Slow Piggy."

ASTROFACT

ONE TRIP AROUND THE SUN TAKES SATURN MORE THAN 29 EARTH YEARS, BUT ONE DAY ON SATURN LASTS JUST UNDER 11 HOURS.

Cassini spacecraft image of Saturn

SPACE DATA

- **DISTANCE FROM THE SUN:** 885.9 MILLION MILES (1.4 BILLION KM)
- **ROTATION PERIOD:** 10.8 HOURS
- **ONE YEAR:** 29.5 EARTH YEARS
- **DIAMETER:** 74,898 MILES (120,536 KM)
- **MOONS:** 62 (53 NAMED)

ASTROFACT

SCIENTISTS BELIEVE THAT SATURN'S RINGS WILL EVENTUALLY DISAPPEAR.

A simulated color view of Saturn's rings

Rings of Ice

Saturn's spectacular rings are made up of billions of icy particles of all different sizes, some as small as grains of sand and others as big as boulders and houses. Scientists aren't quite sure of the rings' exact ages or origins—some may be remnants of a shattered moon or comet. In 1659 Dutch astronomer Christiaan Huygens first theorized that Saturn had a ring around it. Since then the discoveries have kept coming: new rings, smaller ringlets, and the gaps between them. The latest discovery came in 2009 when the Spitzer Space Telescope spotted a dim outermost ring, reaching to perhaps 8 million miles (13 million km) from Saturn itself.

ASTROFACT

EIGHT MILLION MILES (13 MILLION KM) LIE BETWEEN SATURN AND ITS OUTERMOST RING.

Geologically active Enceladus

ASTROFACT

MIMAS'S LONE IMPACT CRATER MAKES THIS MOON OF SATURN'S LOOK LIKE THE DEATH STAR FROM "STAR WARS."

Curious Moons

Many of Saturn's moons exhibit curious features, and at least two, Titan and Enceladus, have the potential for life. Titan's temperature is about minus 289°F (-178°C), much too cold for liquid water but just right for liquid methane. The Cassini spacecraft has found at least two lakes of liquid methane, including the massive Kraken Mare on Titan's northern hemisphere. Titan is the only body other than Earth known to have liquids on its surface. Enceladus's surface is colder still—minus 330°F (-201°C)—so it's too cold for liquids, but underneath its icy crust there could be liquid water. Cassini discovered geysers of water vapor and ice particles whose chemical composition suggests an ocean of liquid water beneath the frigid surface.

ASTROFACT

AT 3,200 MILES (5,150 KM) WIDE, TITAN IS BIGGER THAN THE PLANET MERCURY.

URANUS

Barely visible to the naked eye, the planet Uranus had been erroneously identified as a star by early astronomers before its discovery in 1781. Along with Neptune, Uranus is one of the coldest of the major planets, with temperatures reaching minus 357°F (-216°C). Made largely of water and frozen methane (which lends it a rich turquoise color), the planet is considered an ice giant. Although most images of Uranus show a smooth, opaque ball with few atmospheric markings, in recent years clouds and storms have begun to appear, possibly because the planet is entering its long, warmer spring.

ASTROFACT

IN 1781, WILLIAM HERSCHEL BECAME THE FIRST PERSON TO DISCOVER A PLANET (LATER NAMED URANUS) WITH A TELESCOPE.

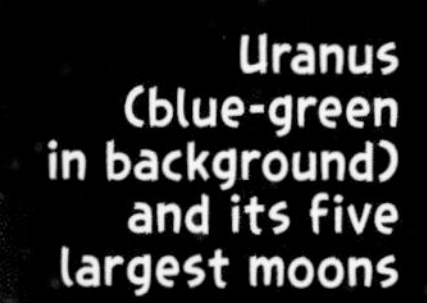

Uranus (blue-green in background) and its five largest moons

ASTROFACT

WINTER LASTS FOR 21 EARTH YEARS ON URANUS.

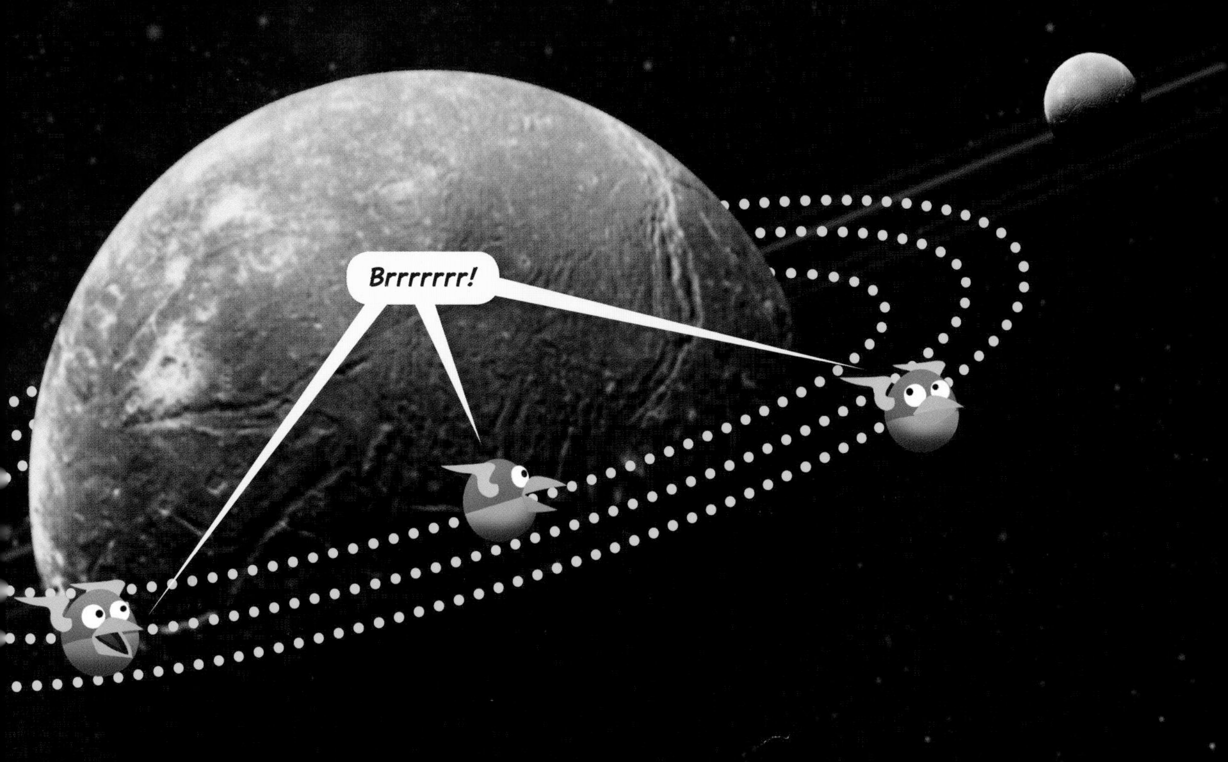

SPACE DATA

- DISTANCE FROM THE SUN: 1.8 BILLION MILES (2.9 BILLION KM)
- ROTATION PERIOD: -17.2 HOURS (RETROGRADE)
- ONE YEAR: 84 EARTH YEARS
- DIAMETER: 31,764 MILES (51,118 KM)
- MOONS: 27

A Tilting Planet

Bland as it may look, Uranus is a little odd. Unlike the other planets that spin more or less perpendicular to the plane of the solar system, Uranus spins on its side. As seasons pass during the course of the Uranian year, first one pole and then (42 Earth years later) the other pole will point to the sun. Its moons—ranging in size from Titania, almost half as big as Earth's moon, to Cordelia, just 16 miles (26 km) wide—circle about the equator flat-on to the sun, so that whenever the planet points one of its poles toward the sun, the Uranian system looks like a bull's-eye. So what knocked Uranus sideways? Some suggest a collision with another planet.

Gentle breezes!

ASTROFACT

WINDS ON URANUS HAVE BEEN CLOCKED AT 360 MILES AN HOUR (579 KM/H).

ASTROFACT

THIRTEEN SLENDER RINGS SURROUND URANUS.

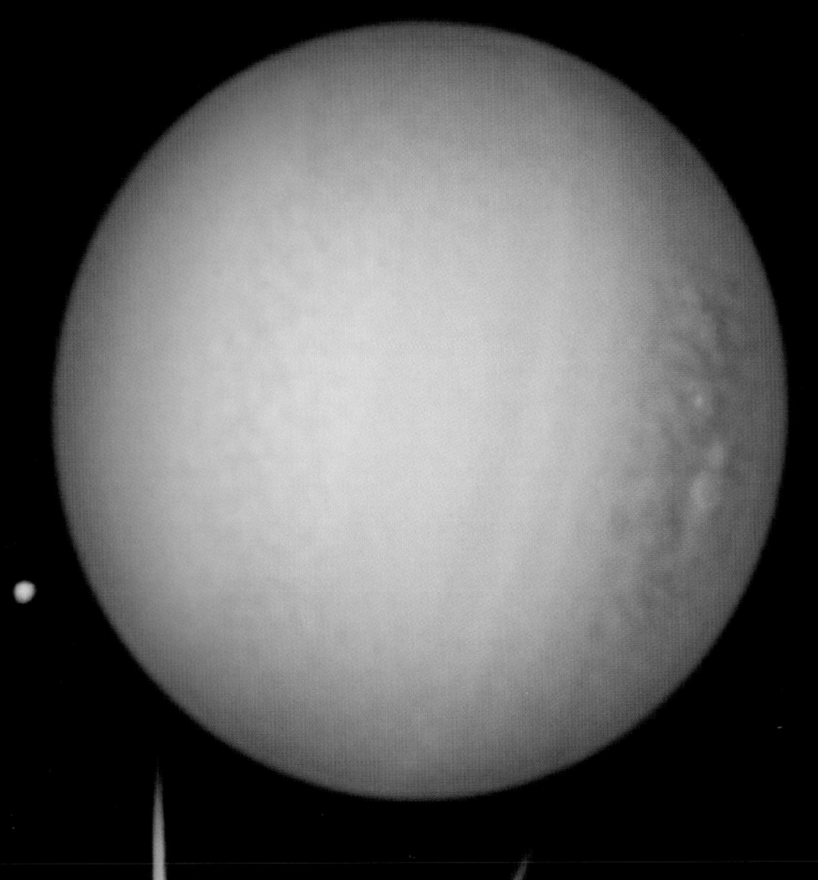

Enhanced color image of Uranus

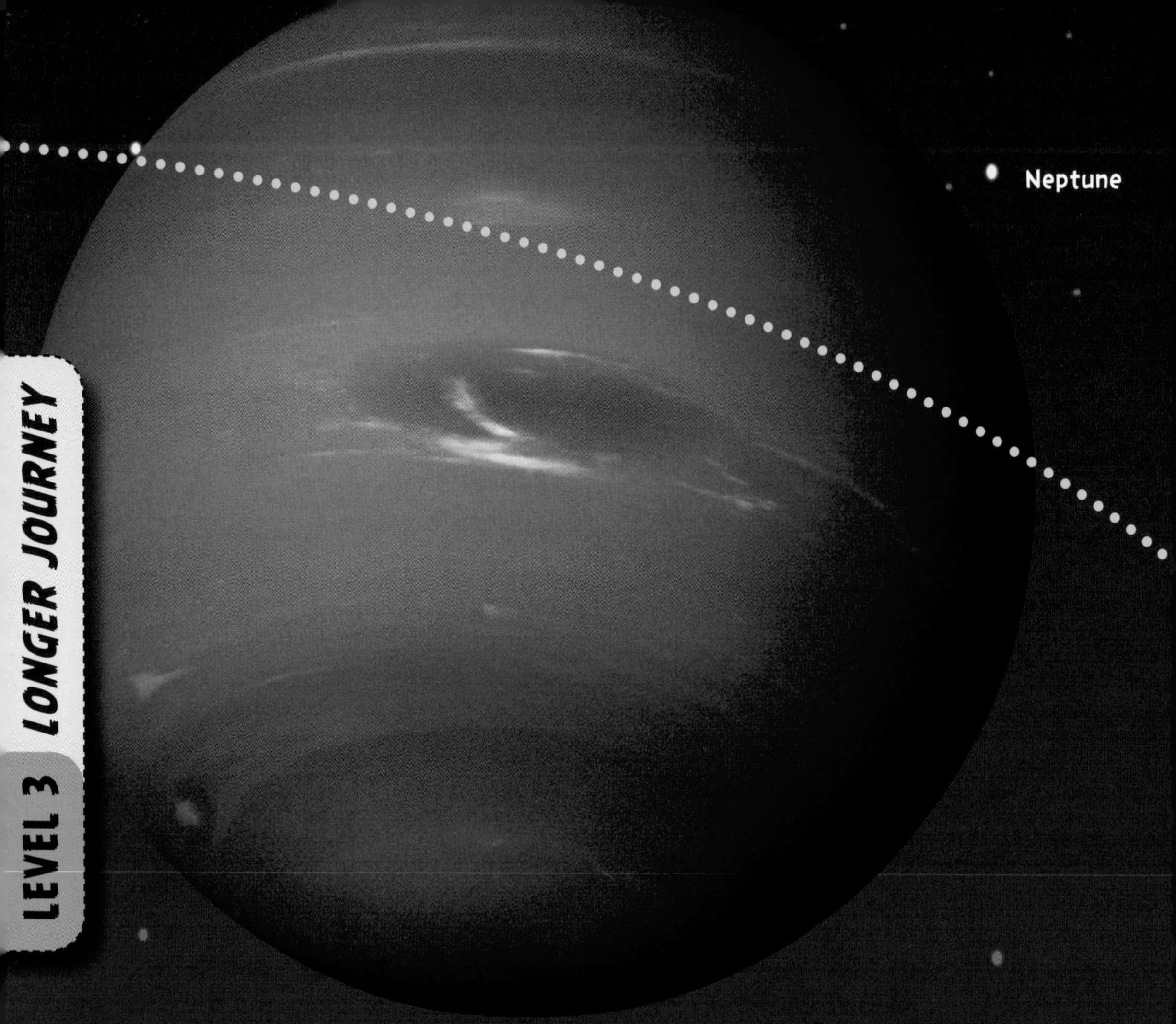

Neptune

ASTROFACT

IF YOU WERE STANDING ON NEPTUNE, YOU WOULD SEE THE SUN AS A DAZZLINGLY BRIGHT STAR IN A BLACK SKY.

NEPTUNE

Distant Neptune is the last major planet in the solar system. An ice giant like Uranus, Neptune is similar in size though somewhat denser. It is enveloped by a deep gas atmosphere composed mainly of hydrogen and helium with a small amount of methane, which gives the planet its vivid blue color. Neptune seems to have an internal heat source, which might explain another Neptunian puzzle: its intensely turbulent weather. Despite receiving little solar energy compared with the inner planets, Neptune has violent storms with supersonic winds of more than 1,200 miles an hour (2,000 km/h), nearly nine times stronger than hurricane winds on Earth.

ASTROFACT

NEPTUNE'S MOON TRITON MAY BE A KIDNAPPED KUIPER BELT OBJECT, CAPTURED BY THE PLANET'S GRAVITY MILLIONS OF YEARS AGO.

SPACE DATA

- **DISTANCE FROM THE SUN:** 2.8 BILLION MILES (4.5 BILLION KM)
- **ROTATION PERIOD:** 16.1 HOURS
- **ONE YEAR:** 164.8 EARTH YEARS
- **DIAMETER:** 30,776 MILES (49,528 KM)
- **MOONS:** 13

VANISHING ACT

Faint rings encircle Neptune . . . for now. First imaged in 1989 by Voyager 2, Neptune's ring system is thin, dusty, and sometimes clumpy and uneven. Astronomers aren't certain of the rings' ages or origins; they speculate that the rings are young and may be the remnants of a former moon torn apart by Neptune's gravity. But the rings could be on their way out. Using the powerful Keck Telescope in Hawaii, astronomers in 2005 observed considerable decay in the rings and believe that they will continue to deteriorate and eventually disappear.

ASTROFACT

NEPTUNE'S RINGS SEEM TO BE MUCH YOUNGER THAN THE PLANET ITSELF.

SOME OF NEPTUNE'S FIVE THIN RINGS ARE LUMPY AND TWISTED.

Neptune's rings as seen by Voyager 2

PLUTO

Earth's inhabitants love little Pluto, and many were very upset when it was demoted in 2006. Pluto was the smallest planet with a diameter of just 1,430 miles (2,301 km). It has four moons of its own and a round shape. But Pluto fails to clear its orbit of debris, and so 76 years after its discovery, Pluto was removed from the planetary ranks. It now lends its name to a new class of solar system objects—plutoids. Pluto and other transneptunian objects inhabit the Kuiper belt, a giant, scattered disk beyond Neptune, containing perhaps a million rocky and icy "leftovers" from the solar system's formation.

ASTROFACT

AN 11-YEAR-OLD GIRL, VENETIA BURNEY, NAMED PLUTO.

Looking back at the sun from Pluto's surface

ASTROFACT

IF YOU TRAVELED AT THE SPEED OF LIGHT, YOU COULD REACH PLUTO IN ABOUT FIVE HOURS.

SPACE DATA

- **DISTANCE FROM THE SUN:** 3.7 BILLION MILES (5.9 BILLION KM)
- **ROTATION PERIOD:** -6.4 HOURS (RETROGRADE)
- **ONE YEAR:** 247.9 EARTH YEARS
- **DIAMETER:** 1,430 MILES (2,301 KM)
- **MOONS:** 4

Pluto and other small worlds located beyond Neptune in the Kuiper belt used to be called dwarf planets, but today they are called plutoids. So far, about a dozen larger, Pluto-like bodies have been discovered. Three of them, Eris, Makemake, and Haumea, have become officially designated plutoids. There are millions of other bodies that revolve around the sun in a broad, flattish ribbon that stretches from Neptune to well past the orbit of Pluto. Astronomers believe that at least 70,000 of them have diameters greater than 60 miles (100 km); hundreds of thousands of smaller ones are invisible to our telescopes, at least for now.

ASTROFACT

POTATO-SHAPED HAUMEA ROTATES ONCE EVERY FOUR HOURS.

ERIS, THE LARGEST KNOWN DWARF PLANET, WAS ORIGINALLY NAMED XENA AFTER TELEVISION'S WARRIOR PRINCESS.

Haumea, a plutoid

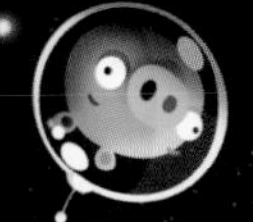

ASTROFACT

A COMET'S NUCLEUS IS MADE OF DIRTY ICE ABOUT AS DARK AS FRESH TAR.

SPACE DATA

NAME: COMET HALLEY

SIZE: 5 X 9.3 MILES (8X15 KM)

ROTATION PERIOD: 2.2 EARTH DAYS

ONE YEAR: 76 EARTH YEARS

NEXT RETURN: 2061

COMETS

As the solar system took shape, billions of balls of frozen gas and dust also formed and were swept out to the farthest reaches. There they stayed, but if any of them are nudged into a new orbit coming closer to the sun, they become comets. Comets warm up as they approach the sun and release stored gases, which form a glowing head, or coma, around the frozen nucleus. A flow of charged ions is shaped by the solar wind into a glowing gas tail. Meanwhile, a stream of dust forms a second tail. Appearances of Halley's comet, one of the most famous, have been recorded since the second century B.C.

ASTROFACT

SUN-GRAZERS ARE COMETS THAT PASS SO CLOSE TO THE SUN THAT THEY EVAPORATE COMPLETELY.

Comet McNaught above the Andes Mountains, January 2007

COMET CLOUD

People once thought that Pluto marked the solar system's outer limit, but studies of long-period comets confounded that theory. In the mid-20th century, astronomers Ernst Öpik and Jan Oort theorized the existence beyond Pluto of an even more distant cloud of icy debris left over from the solar system's formation. Now called the Oort cloud, or sometimes the Öpik-Oort cloud, this collection of comets surrounds the solar system and houses those long-period comets that can take thousands and even millions of years to orbit the sun. While comets with orbital periods less than 200 years originate in the Kuiper belt, the others come from the Oort cloud reservoir.

ASTROFACT

THE OORT CLOUD IS SO DISTANT THAT TELESCOPES HAVE NEVER SEEN IT.

ASTROFACT

THERE ARE UP TO TWO TRILLION ICY ROCKS IN THE OORT CLOUD.

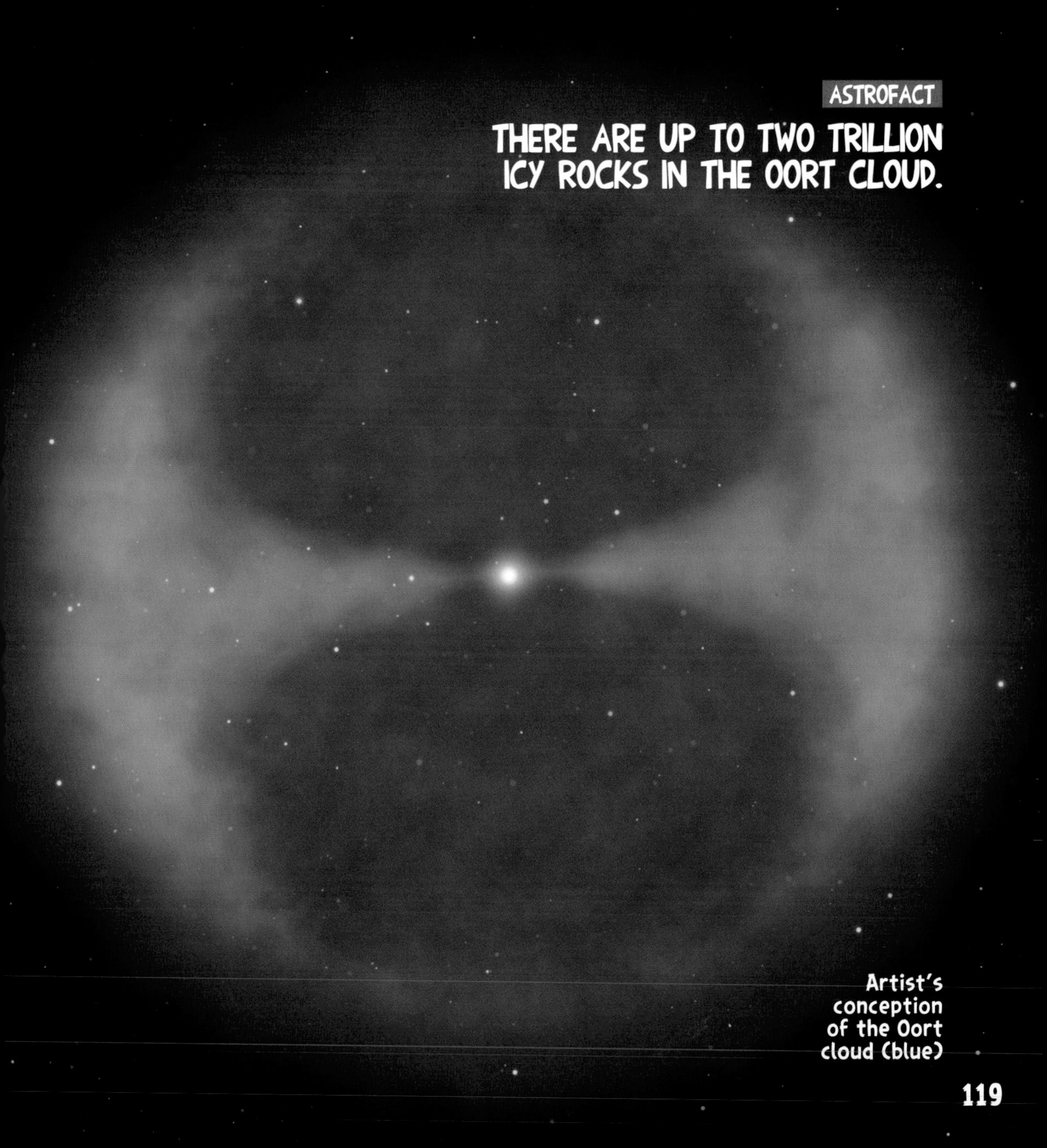

Artist's conception of the Oort cloud (blue)

LEVEL 4 FLYING FARTHER

DEEP SPACE

Massive star Pismis 24-1 illuminates a nebula.

MILKY WAY

For all its great size, despite the trillions of miles from the sun to the distant Oort cloud, our solar system is just one tiny enclave within a vast barred spiral galaxy of several hundred billion stars known to us as the Milky Way. Regions of bright young stars and nebulae crowd the galaxy's spiral arms like traffic jams on freeways looping around a city. Many older stars gradually expel their outer layers, which form beautiful planetary nebulae. Toward the galactic center, a thick swarm of orange and red stars marks the galactic bulge. A corona of old stars and globular star clusters extends far above and below the disk.

IT WOULD TAKE A JUMBO JET ABOUT 120 BILLION YEARS TO FLY ACROSS THE MILKY WAY GALAXY.

Can't that thing move any faster?

ASTROFACT

THE MILKY WAY IS MADE UP OF SOME 200 TO 500 BILLION STARS.

Infrared view of the Milky Way's center

MILKY WAY MIDDLE

Objects in the Milky Way orbit around a galactic center, just as objects in the solar system revolve around the sun. Located about 26,000 light-years from the solar system, the Milky Way's middle is occupied by an object about four million times more massive than the sun. But it isn't a star: It's a gigantic black hole dubbed Sagittarius A*. Despite its size, it doesn't eat very much. Sagittarius A* gets its fuel from winds blown off dozens of massive young stars located a relatively long distance away. The black hole's gravity is powerful, but the fast-moving winds are hard to catch.

ASTROFACT

THE MILKY WAY'S CENTER CONTAINS A SUPERMASSIVE BLACK HOLE.

ASTROFACT

IF YOU LIVED NEAR THE CENTER OF THE MILKY WAY, YOU WOULD FIND THAT MILLIONS OF BRILLIANT STARS KEEP THE SKY BRIGHT ALL NIGHT.

The central stars of the Milky Way galaxy

STARS

The earliest stars formed from condensing hydrogen and helium gases roughly 13.5 billion years ago. Some faded away as white dwarfs, while others, more massive, exploded as supernovae. The tremendous temperatures and pressures involved in these stellar deaths created heavier elements—carbon, nitrogen, oxygen, and others—that were flung into space as each star burst apart in what must have been spectacular light shows. Eventually, the hydrogen gas of the interstellar medium became richer in these heavy elements. And although they represented only one percent of the universe's star stuff, these heavy elements became vital for those of us now living on the rocky planet Earth.

The Pleiades in the constellation Taurus

ASTROFACT

A FEW THOUSAND STARS CAN BE SEEN WITH THE NAKED EYE ON A CLEAR, MOONLESS NIGHT AWAY FROM CITY LIGHTS.

ASTROFACT

SOME SCIENTISTS BELIEVE THAT GOLD IS CREATED WHEN TWO NEUTRON STARS COLLIDE.

SPACE DATA

NEAREST STAR TO THE SUN: PROXIMA CENTAURI

OLDEST KNOWN STAR: HE 1523-0901

LARGEST STAR: VY CANIS MAJORIS

SMALLEST STAR: NEUTRON STARS

MOST MASSIVE STAR: R136A1

STAR SIZE

Neither the biggest nor the smallest, the sun is an average size star in the stellar family. The most common stars are also the smallest—the red dwarfs. These smaller, slower-running stars have such low mass (less than half that of the sun) that they are cooler and able to burn for much longer—up to ten trillion years, by some estimates, which is longer than the age of the universe! Supergiants are the largest stars, like the red supergiant Betelgeuse, which is about 1,000 times bigger than the sun. Found in the constellation Orion, Betelgeuse may be nearing the end of its life and could explode as a supernova in the next few thousand to a million years.

ASTROFACT

RED DWARFS ARE THE MOST COMMON STARS IN THE UNIVERSE.

Blue giants, red giants, and yellow sunlike stars

But are they the angriest?

ASTROFACT

THE HOTTEST STARS SHINE BLUE.

A STAR'S LIFE

Stars are born, they live, and then, when their fuel runs out, they die. How long that takes and how it happens depends on the star's size. As fuel starts to run out, the nuclear core begins to shut down, and the star contracts. Temperatures rise, causing a new round of fusion in which helium atoms are forged into carbon. At this point gravity is overcome by a new surge of energy, and the star begins to expand. Once the available fuel is consumed, the star then sheds its outermost layers of gas, exposing the remains of the core and becoming what is called a white dwarf—a spent, stable, and gradually cooling ember.

Bashing pigs keeps us young!

ASTROFACT

BLUE STRAGGLERS ARE STARS THAT DRAIN FUEL AWAY FROM NEIGHBORING STARS TO STAY "YOUNG."

ASTROFACT

STAR NURSERIES HAVE BEEN FOUND THROUGHOUT THE MILKY WAY IN LARGE GAS CLOUDS.

A cluster of young, hot stars in the Milky Way's Carina arm

ASTROFACT

THE CELLS OF OUR BODIES CONTAIN ELEMENTS MADE IN DISTANT SUPERNOVAE.

The supernova remnant Cassiopeia A in false color

Novae & Supernovae

Novae and supernovae are stars that suddenly become dazzling bright. A nova's brightness varies over time—radiantly glowing over the course of a few hours or days, then slowly fading over the following months. Many novae go through cycles of brightening and dimming, but a supernova is a onetime event, the much brighter result of a dying star. Some supernovae fade quickly, then gradually die. Others first become red supergiants, whose stable cores grow extremely dense. When the core no longer generates energy, it collapses. Material from the star's outer shell rushes toward the core, then rebounds in a massive explosion. A supernova can release so much energy that it may outshine an entire galaxy for days or even weeks.

ASTROFACT

A SUPERNOVA CAN RELEASE (BRIEFLY!) AS MUCH ENERGY AS ALL THE STARS IN THE MILKY WAY COMBINED.

Hmmmm ... How can I tap into that energy?

BLACK HOLES

What happens to a star's core after the violent explosion of a supernova? The dense object left behind either becomes a neutron star—sometimes known as a pulsar for its beaconlike emissions of radiation—or, if large enough, collapses completely and becomes a black hole. The largest supergiants produce such massive gravitational forces that matter literally folds in on itself and collapses to form a black hole. Born in one of the universe's truly strange and awesome events, a black hole is an object in which the distance between subatomic particles has been reduced to zero, and where gravity and density expand toward infinity. Nothing can escape.

SPACE DATA

MAXIMUM MASS FOR A PULSAR: 3 SOLAR MASSES

MINIMUM MASS OF A STELLAR BLACK HOLE: 3 SOLAR MASSES

MINIMUM STELLAR MASS TO BECOME A SUPERNOVA: 1.4 SOLAR MASSES

ASTROFACT

AT THE CENTER OF A BLACK HOLE, THERE IS NO TIME

ASTROFACT

NOTHING, NOT EVEN LIGHT, CAN ESCAPE FROM A BLACK HOLE.

Nothing can escape from me either!

A dust torus surrounding a supermassive black hole

ASTROFACT

PLANETARY NEBULAE HAVE NOTHING TO DO WITH PLANETS.

A section of the Orion Nebula

NEBULAE

There are different kinds of nebulae—giant clouds of gas and dust. Emission nebulae are star-forming clouds set aglow by the energy of the young stars within them. Planetary nebulae are actually the gas that is blown away from a dying red giant, such as in the Ring Nebula in the constellation Lyra. Dark nebulae are collections of interstellar dust that block the light of the stars behind them. Reflection nebulae shine from the light of nearby stars. Nebulae provide the building blocks for stars, galaxies, and planets, and their composition and intricate patterns reveal the nature of both star formation and star death.

ASTROFACT

DARK NEBULAE ARE SO DENSE THAT THEY HIDE ANY STARS WITHIN THEM AS WELL AS THOSE BEHIND THEM.

SPACE DATA

LARGEST: DOZENS OF LIGHT-YEARS ACROSS (DIFFUSE NEBULA)

SMALLEST: ABOUT ONE LIGHT-YEAR ACROSS (PLANETARY NEBULA)

MAXIMUM TIME SPAN: MILLIONS OF YEARS

MINIMUM TIME SPAN: TENS OF THOUSANDS OF YEARS

THE EAGLE NEBULA

Reminds me of a mighty friend of mine . . .

The Eagle Nebula is a stellar nursery in the Milky Way. Stars form in the clouds of hydrogen gas and produce ultraviolet light, causing the gas clouds to glow with color. Located in the constellation Serpens, about 7,000 light-years away, the Eagle Nebula was discovered independently by Philippe Loys de Chéseaux between 1745 and 1746 and then by Charles Messier in 1764. Messier described the stars in the nebula as "enmeshed in a faint glow." The nebula has dark pillars of dense material rising in its center; these can be seen with a 12-inch telescope.

ASTROFACT

THE DARK TOWER OF GAS RISING FROM THE EAGLE NEBULA IS 56 TRILLION MILES (90 TRILLION KM) HIGH.

The Eagle Nebula
ASTROFACT
IN THE EAGLE NEBULA, STARS ARE BORN FROM SOLAR SYSTEM-SIZE CLUMPS OF COLD GAS.

INSIDE THE SWAN

ASTROFACT

SHOCK WAVES FROM THE ORIGINAL SUPERNOVA MOVED AT 370,000 MILES AN HOUR (600,000 KM/H).

The constellation Cygnus the Swan spans one of the most interesting parts of the night sky, an area of the Milky Way that is packed with bright stars and fascinating celestial objects. One of them, the Veil Nebula, is the remnant of a supernova, an explosion that occurred 15,000 years ago and left a macramé-like nebula threaded with knots and filaments. The nebula is so large that several sections have been given their own astronomical designations. One section, called the Witch's Broom Nebula, flies near the glowing star 52 Cygnus.

ASTROFACT

THE WITCH'S BROOM NEBULA IS PART OF THE CLOUD LEFT OVER FROM A SUPERNOVA EXPLOSION ABOUT 1,400 LIGHT-YEARS AWAY.

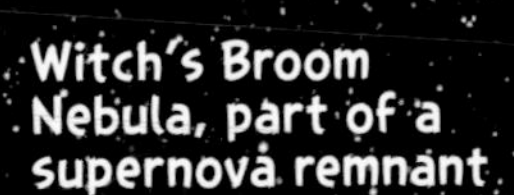

Witch's Broom Nebula, part of a supernova remnant

The spiral galaxy M74

GALAXIES

Galaxies come in three primary shapes: elliptical, spiral—which are large and massive—and irregular, which are small. Elliptical galaxies have almost no visible internal structure, contain little gas and dust, and typically contain older red giant stars. Spiral galaxies come in two types: basic and barred. Spiral galaxies hold a mixture of old and young, hot and cool stars. They also hold gas and dust that promote star formation. Barred spirals, like the Milky Way, have an elongated nuclear bulge that looks like a bar of stars cutting through the galaxy's middle. Like spiral galaxies, irregular galaxies hold a variety of star types and large clouds of gas and dust, but have no distinct structure.

ASTROFACT

GALAXIES SEEM TO BE SURROUNDED BY INVISIBLE CLOUDS OF DARK MATTER, TEN TIMES AS MASSIVE AS ALL THEIR STARS.

ASTROFACT

125 BILLION GALAXIES ARE ARRAYED THROUGHOUT THE UNIVERSE.

Galaxies Collide

Galaxies tend to cluster in groups ranging from two or three members to as many as a few thousand. Our own galaxy resides in a community of about 50, known as the Local Group. Galaxies do interact and may be caught in a gravitational embrace that often results in collisions and galactic mergers that trigger star formation. Occasionally the interaction creates long tails of stars and gas that stream from one galaxy to the other, forming a bridge. Sometimes a giant galaxy—like the Milky Way—with its immense gravity, rips apart a smaller galaxy nearby. The enormous galactic "cannibal" eventually devours the pieces of its neighbor left behind.

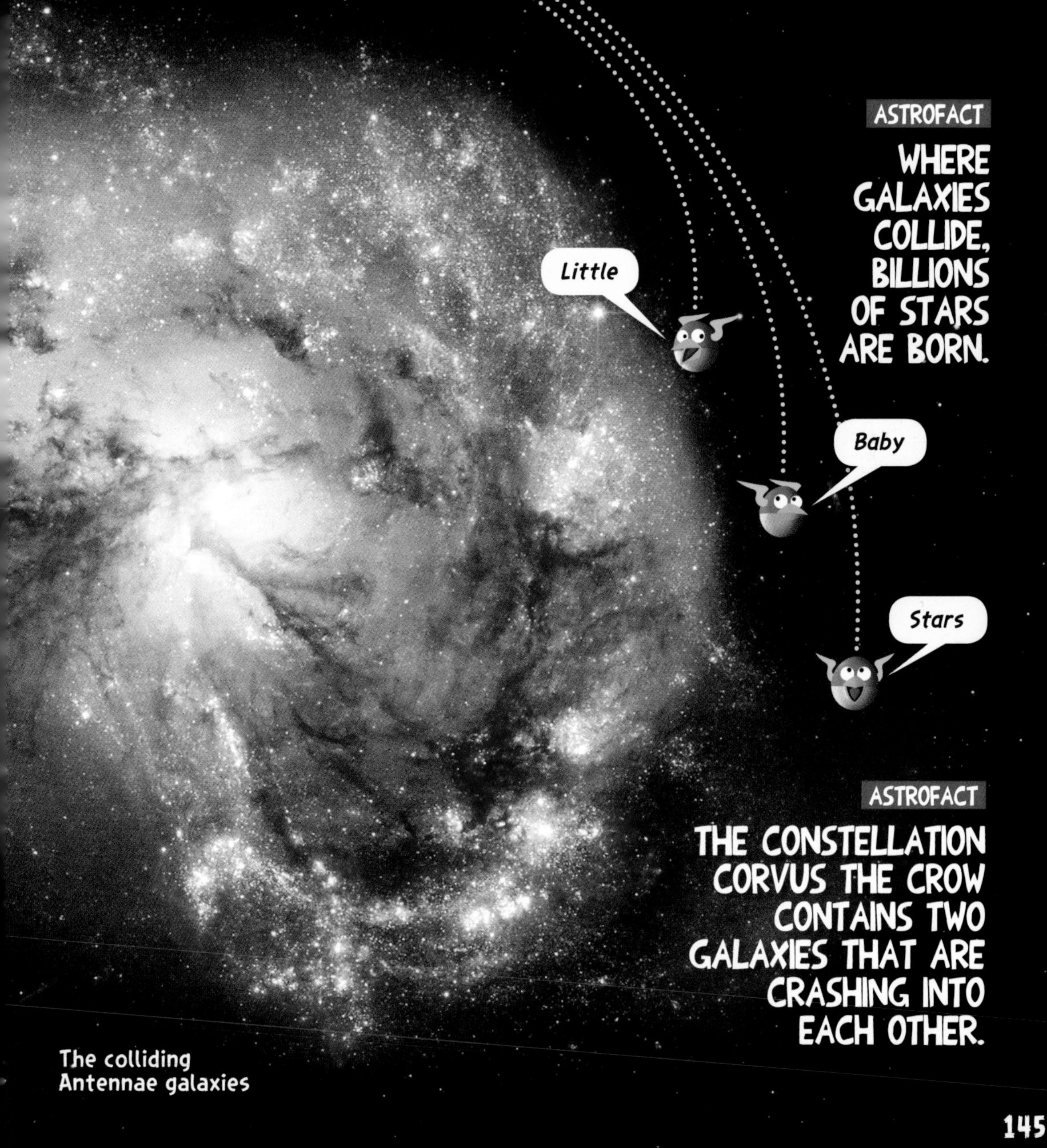

The colliding Antennae galaxies

ASTROFACT

WHERE GALAXIES COLLIDE, BILLIONS OF STARS ARE BORN.

ASTROFACT

THE CONSTELLATION CORVUS THE CROW CONTAINS TWO GALAXIES THAT ARE CRASHING INTO EACH OTHER.

EXOPLANETS

The hunt for exoplanets—planets orbiting other stars—is one of the next great goals of astronomers. The first extrasolar planet was discovered in 1991 by radio astronomers. Since then, the astronomical toolbox has expanded to include advanced ground-based telescopes, like Hawaii's Keck Observatory, and space-based equipment. In 2008, the Hubble Space Telescope returned the first visual images of an exoplanet, Fomalhaut b, named for its parent star. The Spitzer Space Telescope has found gas-rich exoplanets with organic molecules. Since 2009, the Kepler Space Telescope has been scanning the Milky Way for planets and has found more than 2,300 planetary candidates.

ASTROFACT

LOCATED ABOUT 750 LIGHT-YEARS FROM EARTH IS THE MILKY WAY'S DARKEST PLANET. ITS COAL-BLACK SURFACE REFLECTS ALMOST NO LIGHT.

Artist's depiction of a Jupiter-like exoplanet and its parent star, Fomalhaut

ASTROFACT

IN 2011, THE KEPLER SPACE TELESCOPE DISCOVERED THE FIRST PLANET KNOWN TO ORBIT A PAIR OF STARS.

SPACE DATA

LONGEST KNOWN "YEAR": FOMALHAUT B, ABOUT 900 YEARS

FIRST DISCOVERED: PSR 1257+12 B, IN 1991

SHORTEST KNOWN "YEAR": 55 CANCRI E, 17 HOURS, 41 MINUTES

ANOTHER EARTH?

Since 2009, the Kepler Space Telescope has found more than 2,300 planet candidates (most the size of Neptune and larger) but the search for another Earth was fruitless at first—planets were too hot or too cold for water to be liquid. But in 2011, NASA found something exciting orbiting a star about 600 light-years away: a planet they call Kepler 22-b. This world exists in the "habitable zone" and has a pleasant surface temperature of 72°F (22°C). Scientists don't know yet if the planet is rocky, gaseous, or liquid, but it's the most promising candidate yet for life beyond our solar system.

ASTROFACT

ABOUT 10 PERCENT OF ALL STARS MAY HAVE EARTH-SIZE WORLDS.

ASTROFACT

ASTRONOMERS CALL THE IDEAL EARTHLIKE WORLD A "GOLDILOCKS PLANET"—NOT TOO HOT, NOT TOO COLD, BUT JUST RIGHT.

Artist's conception of planet Kepler 22-b

THE DARK SIDE

Cosmologists believe that following the big bang, the cosmos began expanding. Believing that gravity was slowing the expansion, research teams in the mid-1990s were surprised to discover that expansion was speeding up. If the universe is expanding faster, they theorized that a mysterious repulsive force—dark energy—must be pushing it apart. Recent observations indicate that the some 70 percent of the cosmos is made of dark energy, but scientists aren't sure how it behaves. Dark energy's influence could dictate the ultimate fate of the universe—it may continue to expand and even accelerate to speeds that rip the cosmos apart. But the universe could gradually decelerate, contracting into a "big crunch"—perhaps only to expand again.

ASTROFACT

SOME PHYSICISTS BELIEVE THAT OUR UNIVERSE IS JUST ONE OF AN ALMOST INFINITE NUMBER IN A "MULTIVERSE."

ASTROFACT

DARK ENERGY IS BELIEVED TO THWART GRAVITY.

Hot gas and dark matter in Pandora's cluster

SPACE DATA

HOW OLD IS THE UNIVERSE? 13.7 BILLION YEARS

WHEN DID THE FIRST STARS FORM? 150 MILLION YEARS AFTER THE BIG BANG

WHEN DID THE FIRST GALAXIES FORM? 500 MILLION YEARS AFTER THE BIG BANG

Glossary

Asteroid: small, rocky body revolving around the sun; the majority of asteroids orbit between Mars and Jupiter in the asteroid belt.

Astrobiology: the study of life in the universe and the search for extraterrestrial life.

Astronomical unit (AU): the average distance from Earth to the sun, 92,955,730 miles (149,597,870 km); employed as a standard unit of astronomical measurement.

Big bang model: the widely accepted theory of the origin and evolution of the universe that states the universe began in an infinitely compact state and is now expanding.

Brown dwarf: an object between a planet and a star in size but lacking enough mass to start nuclear fusion in its core.

Comet: a small, icy body orbiting the sun; short-period comets originate in the Kuiper belt, while long-period comets originate in the Oort cloud.

Corona: the sun's outermost atmosphere; the visible light from the sun, seen around the disk of the moon as it covers the sun during a total eclipse.

Coronal mass ejections: bubble-shaped bursts of electrically charged gas from the solar corona; often associated with large flares.

Dwarf planets: objects orbiting the sun that are large enough to become round but have not swept up all the debris in their orbital paths.

Dwarf stars: stars of roughly the sun's mass and less; the category includes white and red dwarf stars.

Eclipse: the partial or total covering of one celestial body by another, such as the sun by the moon.

Exoplanet: extrasolar planet; a planet orbiting a star other than the sun.

Extremophile: an organism that lives in extreme environmental conditions.

Galaxy: a collection of billions of stars held together by their mutual gravity.

Gas giant: a large, low-density planet composed primarily of hydrogen and helium; in our solar system Jupiter, Saturn, Uranus, and Neptune are considered gas giants; Uranus and Neptune are also called ice giants.

Gravity: the attractive force that every object in the universe has on every other; also called gravitation.

Habitable zone: the orbital distance from a star where water can remain on a planet's surface; Earth orbits within the sun's habitable zone.

Inner planets: the four small, rocky planets nearest to the sun; also called terrestrial planets.

Interstellar space: the region between stars.

Kuiper belt: (pronounced KYE-per) a disk-shaped region located 30 to 50 AU from the sun that is filled with icy bodies; the source of most short-period comets.

Light-year: the distance light travels in a vacuum in one year; equivalent to about 6 trillion miles (9.5 trillion km).

Meteor: the streak of light produced by an object entering Earth's atmosphere and traveling so fast that friction causes it to vaporize.

Milky Way: the large spiral galaxy that contains more than 200 billion stars and is home to the sun.

Moon: a body in orbit around a planet, dwarf planet, or asteroid; a natural satellite.

Near-Earth object (NEO): an asteroid or short-period comet whose orbit brings it into the Earth's orbital neighborhood, where it could collide with our planet.

Nebula: a cloud of gas and dust in interstellar space.

Oort cloud: an enormous spherical cloud of icy objects surrounding our solar system at distances up to 100,000 AU; thought to be the region where comets originate.

Orbit: the path followed by an object moving in a gravitation field of a larger object, as in the path of a planet around the sun or a moon around a planet.

Planet: a body in orbit around the sun, possessing sufficient mass for its self-gravity to bring it to a nearly round shape, and gravitationally dominant, meaning it will have cleared its path of other bodies of comparable size, other than its satellites.

Planetary nebula: a shell of ejected material moving away from an extremely hot, dying star.

Planetary rings: billions of pieces of rock and ice organized into thin, flat rings orbiting planets.

Plutoid: dwarf planets in orbit around the sun outside the orbit of Neptune.

Red giant: a star that has used up its hydrogen core and expanded to more than 100 times its original size.

Solar flares: sudden and intense outbursts of the lower layers of the sun's atmosphere, usually near a sunspot group.

Solar wind: an outflow of charged particles from the sun.

Stellar nursery: dense clouds of interstellar gas and dust in which stars are born.

Sunspot: a large, dark, relatively cool spot on the surface of the sun, associated with strong magnetic activity.

Supergiant star: a member of a class of the largest and most luminous stars known, with diameters between 100 and 1,000 times that of the sun.

Supernova: the explosive death of a star that has at least several times the sun's mass. A supernova may temporarily equal an entire galaxy in brightness.

White dwarf: a star that has burned up all of its nuclear fuel and has collapsed to a fraction of its former size while still retaining a significant mass.

Sky-watching Appendix

The sky above is full of amazing sights—just look up. Here's a short list of the things you can see from your own backyard.

Be sure to consult astronomy magazines and other resources for exact viewing times. All eclipse data by Fred Espenak, NASA's GSFC.

TOTAL SOLAR ECLIPSES

Year	Date	Viewing Location
2012	November 13	Northern Australia
2015	March 20	Iceland, northern Europe, North Africa, north Asia
2016	March 9	Sumatra, Borneo, Sulawesi
2017	August 21	North America, northern South America
2019	July 2	Chile, Argentina
2020	December 14	Chile, Argentina

PARTIAL SOLAR ECLIPSES

Year	Date	Type	Viewing Location
2012	May 20	Annular	China, Japan, Pacific Islands, western United States
2013	May 10	Annular	Northern Australia, Solomon Islands
2014	April 29	Annular	Australia, Antarctica
2014	October 23	Partial	Western North America
2015	September 13	Partial	South Africa, Antarctica
2016	September 1	Annular	Africa, Indian Ocean
2017	February 26	Annular	Chile, Argentina, Africa, Antarctica
2018	February 15	Partial	Southern South America, Antarctica
2018	July 13	Partial	Southern Australia
2018	August 11	Partial	Northern Europe, northeast Asia
2019	January 6	Partial	Northeast Asia, northern Pacific
2019	December 26	Annular	Saudi Arabia, India, Sumatra, Borneo, Australia
2020	June 21	Annular	Central Africa, Southern Asia, China

TOTAL LUNAR ECLIPSES

Year	Date	Viewing Location
2014	April 15	Australia, Pacific Islands,western North America
2014	October 8	Australia, Pacific Islands, western North America
2015	April 4	Australia, Pacific Islands, western North America
2015	September 28	Europe, Africa, western Asia, eastern Pacific, Americas
2018	January 31	Asia, Australia, Pacific Islands, western North America
2018	July 27	South America, Europe, Africa, Asia, Australia
2019	January 21	Central Pacific, Americas, Europe, Africa

Peak activity can vary, so consult resources for exact viewing times.

MAJOR ANNUAL METEOR SHOWERS

Shower	Constellation	Dates of Activity	Best Views
Quadrantids	Draco	Jan. 1-5	Northern Hemisphere
Alpha Centaurids	Centaurus	Feb. 6-8	Southern Hemisphere
Gamma Normids	Norma	Mar. 12-14	Southern Hemisphere
Eta Aquarids	Aquarius	Apr. 19-May 28	Northern Hemisphere
Pi Puppids	Puppis	Apr. 22-24	Southern Hemisphere
Perseids	Perseus	July 17-Aug. 24	Northern Hemisphere
Orionids	Orion	Sept. 10-Oct. 26	Northern Hemisphere
Leonids	Leo	Nov. 14-21	Northern Hemisphere
Geminids	Gemini	Dec. 7-17	Northern Hemisphere

COMETS

Comet	Orbit Period	Next Date Visible
1P/Halley	76 years	2061
109P/Swift-Tuttle	130 years	2127
C/1995 O1 Hale-Bopp	2,400 years	4397

Further Information

National and International Organizations

British Astronomical Association, britastro.org/baa
Website for amateur astronomers in the United Kingdom

European Space Agency, www.esa.com
Official website for the European Space Agency (ESA)

International Astronomical Union, www.iau.org
Official website for the International Astronomical Union

National Aeronautics and Space Administration, www.nasa.gov
Official website for the National Aeronautics and Space Administration (NASA)

Space Weather Prediction Center, www.swpc.noaa.gov
Website run by the National Oceanic and Atmospheric Administration, which reports on the space environment surrounding Earth

Observatories and Telescopes

Chandra X-Ray Observatory, chandra.harvard.edu
Official website for the Chandra, the first space-based x-ray observatory

European Southern Observatory, www.eso.org/public
Official website for the European Southern Observatory, which conducts astronomical research in the Southern Hemisphere

Hubble Site, hubblesite.org/newscenter
The official website for the Hubble Space Telescope

Mauna Kea Observatories, www.ifa.hawaii.edu/mko
Official website for the world's largest observatory for infrared, optical, and submillimeter astronomy

National Optical Astronomy Observatory, www.noao.edu
Official website for Kitt Peak National Observatory, Cerro Tololo Inter-American Observatory, and National Optical Astronomy Observatory (NOAO) Gemini Science Center

Palomar Observatory, www.astro.caltech.edu/Palomar
Official website for the Palomar Observatory in San Diego, California

W.M. Keck Observatory, www.keckobservatory.org
Website for the twin Keck Telescopes, the largest optical and infrared telescopes in the world

Books

Aguilar, David A. *13 Planets: The Latest View of the Solar System.* Washington, D.C.: National Geographic Society, 2011.

Daniels, Patricia. *My First Pocket Guide: Constellations.* Washington, D.C.: National Geographic Society, 2002.

Daniels, Patricia. *The New Solar System.* Washington, D.C.: National Geographic Society, 2009.

Dickinson, Terence. *Night Watch: A Practical Guide to Viewing the Universe.* Buffalo: Firefly Books, 2006.

Glover, Linda K. *National Geographic Encyclopedia of Space.* Washington, D.C.: National Geographic Society, 2005.

Lang, Kenneth R. *The Cambridge Guide to the Solar System.* Cambridge: Cambridge University Press, 2003.

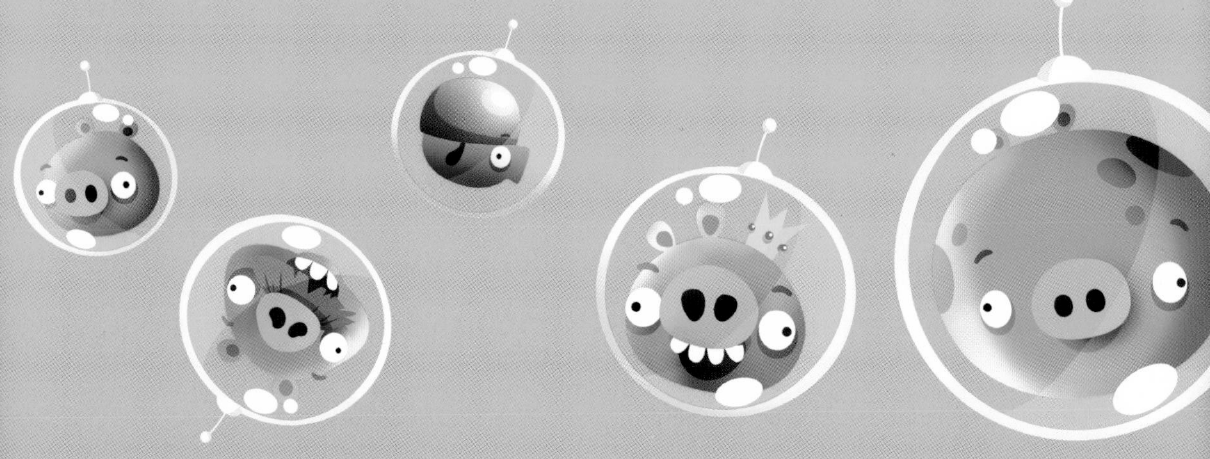

Illustration Credits

Cover, artwork by NASA/ESA and Adolf Schaller; 1, NASA/CORBIS; 2-3, NASA/ESA/H. E. Bond, STScI; 8-9, NG Maps; 10-11, Siddhartha Saha; 12-13, artwork by David Aguilar; 14-15, artwork by David Aguilar; 17, Mary Evans Picture Library; 18-19, Mark Thiessen; 20, NASA; 23, NASA; 24, NASA; 26, NASA; 26-27, NASA; 28, artwork by NASA/JPL-Caltech; 30-31, artwork by NASA/JPL; 32-33, artwork by Johns Hopkins University Applied Physics Laboratory/Southwest Research Institute; 34-35, NASA; 36-37, NASA/JPL-Caltech; 38, NASA/CXC/PSU/L.Townsley et al.; 41, Seth Shostak/SETI Institute; 42-43, artwork by SPL/Photo Researchers, Inc.; 44-45, SOHO/ESA/NASA; 46-47, artwork courtesy NASA GSFC; 48-49, Oddbjorn Engvold, Jun Elin Wiik, Luc Rouppe Van Der Voort, Institute of Theoretical Physics, University of Oslo Swedish 1-Meter Solar Telescope; 50, Markus J. Aschwanden, Lockheed Martin Solar and Astrophysics Laboratory (LMSAL), recorded in extreme ultraviolet from NASA's transition region and coronal explorer (TRACE) satellite; 52-53, artwork by NASA/Johns Hopkins University Applied Physics Laboratory/Carnegie Institution of Washington; 54, NASA/Johns Hopkins University Applied Physics Laboratory/Arizona State University/Carnegie Institution of Washington. Image reproduced courtesy of Science/AAAS; 56-57, David P. Anderson, Southern Methodist University/NASA/JPL; 58, Terrance Emerson/Shutterstock; 60-61, artwork by photovideostock/iStock; 62-63, Pacific Ring of Fire 2004 Expedition, NOAA Office of Ocean Exploration; Dr. Bob Embley, NOAA PMEI, Chief Scientist; 64-65, Stephen Alvarez/National Geographic Stock; 67, NASA/JPL/USGS; 68-69, jwohlfeil/Shutterstock; 70-71, artwork by Fahad Sulehria, www.novacelestia.com; 72, artwork courtesy NASA; 75, NASA/USGS; 76-77, NASA/JPL/Cornell University; 78-79, NASA/JPL/University of Arizona; 80, Mary Evans Picture Library; 80-81, Anthony Fiala; 82-83, artwork by Roger Harris/Photo Researchers, Inc.; 84-85, artwork by David Aguilar; 86-87, artwork by Sanford/Agliolo/CORBIS; 88-89, NASA/JPL/Space Science Institute; 90, NASA/JPL; 93, NASA/M. Wong and I. de Pater, University of California, Berkeley; 94, NASA/JPL/DLR; 96, NASA/JPL/University of Arizona/University of Colorado; 98-99, NASA/JPL/Space Science Institute; 100-1, NASA/JPL; 102, NASA/JPL/Space Science Institute; 104-105, artwork by Mark Garlick/Photo Researchers, Inc.; 107, Science Source; 108, NASA/JPL; 110-111, Science Source; 112-113, artwork by John R. Foster/Photo Researchers, Inc.; 114-115, artwork by David Aguilar; 116-117, Miloslav Druckmoller; 119, artwork by Mark Garlick/Photo Researchers, Inc.; 120-121, NASA/ESA/Jesús Maíz Apellániz, Instituto de Astrofísica de Andalucía, Spain; 122-123, M. Hauser STScI, the COBE/DIRBE Science Team, and NASA; 125, ESO/S. Gillessen et al.; 126-127, NASA/ESA and AURA/Cal; 128, NASA/ESA/Hubble/SM4 ERO Team; 130-131, NASA/ESA/Hubble Heritage STScI/AURA-ESA/Hubble Collaboration; 132, NASA/JPL-Caltech/O. Krause, Steward Observatory; 134-135, artwork by ESA/NASA, the AVO project and Paolo Padovani; 136-137, NASA/ESA/Hubble Heritage Team STScI/AURA; 139, T. A. Rector and B. A. Wolpa/NRAO/AUI/NSF; 140-141, T. A. Rector, U. Alaska/WIYN/NOAO/AURA/NSF; 142-143, NASA/ESA; 144-145, NASA/ESA/Hubble Heritage STScI/AURA-ESA/Hubble Collaboration; 146-147, artwork by ESA/NASA/L. Calcada, ESO for STScI; 148-149, artwork by NASA/Ames/JPL-Caltech; 150-151, X-ray, NASA/CXC/ITA/INAF/J. Merten et al; Lensing, NASA/STScI, NAOJ/Subaru, ESO/VLT; Optical, NASA/STScI/R. Dupke.

Acknowledgments

We would like to extend our thanks to the terrific team who worked so hard to make this project come together so quickly and so well.

Rovio

Sanna Lukander, Antti Grönlund, Hanna Silvennoinen, Jan Schulte-Tigges

National Geographic

Amy Briggs, Jonathan Halling, Susan Blair, Judith Klein, Joan Gossett, Rob Waymouth, Anna Zusman, Lisa A. Walker, Robert Burnham, and Patricia Daniels